ANIMAL MEMORY

CONTRIBUTORS

BERNARD W. AGRANOFF

E. J. CAPALDI

RONALD G. DAWSON

HENRY GLEITMAN

JAMES L. MCGAUGH

SAM REVUSKY

NORMAN E. SPEAR

EUGENE WINOGRAD

ANIMAL MEMORY

Edited by Werner K. Honig and P. H. R. James

Department of Psychology
Dalhousie University
Halifax, Nova Scotia

ACADEMIC PRESS   New York and London   1971

ACADEMIC PRESS, INC.
111 Fifth Avenue, New York, New York 10003

United Kingdom Edition published by
ACADEMIC PRESS, INC. (LONDON) LTD.
24/28 Oval Road, London NW1 7DD

LIBRARY OF CONGRESS CATALOG CARD NUMBER: 74-159620

PRINTED IN THE UNITED STATES OF AMERICA

CONTENTS

Chapter 4.   **The Role of Interference in Association over a Delay**

Sam Revusky

Chapter 5.   **Modification of Memory Storage Processes**

James L. McGaugh and Ronald G. Dawson

Chapter 6.   **Effects of Antibiotics on Long-Term Memory Formation in the Goldfish**

Bernard W. Agranoff

Chapter 7. Some Issues Relating Animal Memory to Human Memory

Eugene Winograd

LIST OF CONTRIBUTORS

Numbers in parentheses indicate the pages on which the authors' contributions begin.

BERNARD W. AGRANOFF (243), Mental Health Research Institute and Department of Biological Chemistry, University of Michigan, Ann Arbor, Michigan

E. J. CAPALDI (111), Department of Psychology, Purdue University, Lafayette, Indiana

RONALD G. DAWSON (215), Department of Psychobiology, School of Biological Sciences, University of California, Irvine, California

HENRY GLEITMAN (1), Department of Psychology, University of Pennsylvania, Philadelphia, Pennsylvania

JAMES L. MCGAUGH (215), Department of Psychobiology, School of Biological Sciences, University of California, Irvine, California

SAM REVUSKY* (155), Department of Psychology, Northern Illinois University, DeKalb, Illinois

NORMAN E. SPEAR (45), Department of Psychology, Rutgers University, New Brunswick, New Jersey

EUGENE WINOGRAD (259), Department of Psychology, Emory University, Atlanta, Georgia

* Present address: Department of Psychology, Memorial University of Newfoundland, St. John's, Newfoundland, Canada.

PREFACE

After a long period of neglect, experimental and theoretical work in the area of animal memory has recently revived. The reasons for the neglect may be historical, since the first systematic experimental work on forgetting was carried out with (and by) a human subject, Hermann Ebbinghaus. They may be due to a concentration of effort on acquisition processes in animal learning rather than forgetting. Finally, the neglect may be due to the fact that "memory" is often reified as a psychological entity which would be more easily examined through human verbal report rather than animal behavior. But the advantages of using animal subjects to study forgetting and retention are obvious. It is much easier to control the extraexperimental sources of interference in animals than in humans, both before and after the behavior of interest is initially acquired. Animals can be held for long retention intervals which would normally involve a considerable loss within a sample of human subjects. Such physiological treatments as electroconvulsive shock and the administration of drugs that affect the brain can be carried out on animals after acquisition. Finally, there is no reason to believe that processes of retention, recall, and forgetting are any less firmly based on empirical observation than the process of acquisition.

These observations are clearly illustrated in this survey of current work in the area of animal memory. Since the contents of the book are based on a symposium, coverage is selective rather than comprehensive. However, each chapter provides a broad coverage of the topic with which it is concerned, and the experimental work reported here is certainly representative of the most significant developments in the field.

Two chapters are concerned with what we may call "associative memory"—the memory of one event which is essential to its association (over a delay) with subsequent events. Capaldi shows clearly that animals can remember events from one learning trial to the next and that their behavior will be determined largely by the sequences of trials with differing outcomes. The bearing of these associative processes on phenomena such as the partial reinforcement effect and alternation is clearly discussed. Revusky reports a great deal of research on the association of flavors with toxicosis in a conditioning paradigm, but with long delays between the conditioned and unconditioned stimuli. He presents a theory of association and forgetting which is based largely on concepts of interference.

The chapters by Gleitman and by Spear examine the current situation with respect to what we may call "retentive memory"—i.e., the retention and forgetting of learned behavior over time. Gleitman argues strongly that animals do forget, and examines theories of forgetting, with special attention to interference theories, which do not seem to adequately explain the phenomenon. Spear also develops a theoretical position, supported by extensive data on animal forgetting. Distinguishing between "lapses" and "losses," he argues in favor of a retrieval theory of remembering. Much forgetting is to be explained by the absence of critical stimuli, or "cues," at the time of recall, rather than a permanent loss from the memory "store."

The chapters by McGaugh and by Agranoff examine the physiological basis of memory in terms of consolidation theory. Electroconvulsive shock and biochemical injections, administered at various times after initial acquisition, indicate that a process of "consolidation," perhaps involving transfer from short- to long-term memory, is subject to interference. The complex paradigms required to support such a notion are reviewed, and theoretical models are proposed by each author. Finally, Winograd provides a critical discussion based on all of the foregoing material in which the topics covered in the book are related in an informative and constructive way to current work on human retention and forgetting.

This book is based on the proceedings of a symposium held at Dalhousie University in the summer of 1969. The chapters have been revised and the material has been brought up-to-date since that time. We are very grateful to the Faculty of Graduate Studies of Dalhousie University for generous financial support which made the symposium possible. Our colleagues in the Psychology Department participated in organizing and directing the symposium. Dr. and Mrs. C. J. Brimer carried a particularly heavy share of looking after practical arrangements. Finally, we want to thank Miss Margaret Ross, who has done extensive secretarial and editorial work on the preparation of the contributed papers for publication. Without her services at a time when both editors were absent from Dalhousie for long periods of time, the publication of this book would barely have been possible.

WERNER K. HONIG
P. H. R. JAMES

ANIMAL MEMORY

Chapter I

FORGETTING OF LONG-TERM
MEMORIES IN ANIMALS[1]

Henry Gleitman

The past two decades have seen an enormous burgeoning of interest in the field of memory. Much of the research has been undertaken in the hope of getting at the physiological basis of the memory trace, and it is hardly surprising that most of these investigations used animal subjects. At least thus far, much of this effort has concentrated upon the early stages in the life of the memory trace. The biological basis of trace formation, decay and consolidation of short-term memory, disruptions of the consolidation process, transfer from the short-term to the long-term store—these are some of the topics that by now constitute a substantial portion of the research literature. But interestingly enough, there has been no comparable concern with the later phase in the life of the trace. Once formed, consolidated, and transferred to long-term storage, the memory trace seemed to hold little further interest (at least until fairly recently) to investigators working at the animal level, despite the fact that forgetting of long-term memories is such an obvious fact of our own human experience.

Of course, this is not to say that forgetting of long-term memories has not been studied seriously. Quite the contrary. Research on human forgetting dates back to 1885 (Ebbinghaus, 1885) and ever since, the phenomenon has been a primary concern of psychologists with a functionalist orientation (e.g., McGeoch, 1942) who have dealt with it in innumerable investigations of human rote learning. But for the interpretation of their data, oddly enough, they often turned to a realm quite different from that which had supplied the data they wanted to interpret. The data (both for forgetting and for what were thought to be the allied effects of retroactive and proactive interference) stemmed almost exclusively from studies of human verbal and motor learning. On the other hand, the inferred mechanisms offered as explanations for these data were usually

[1] This research was supported by U.S. Public Health Service Research Grants M-4993 and MH-10629.

derived from phenomena observed primarily at the level of conditioning and animal learning. Thus Underwood and Postman (1960) explained forgetting as response competition caused in large part by the spontaneous recovery of inappropriate habits extinguished in the course of original learning. There was a gap here, which was all the more obvious in view of the relative paucity of research on animal forgetting. When one talks of bar-pressing and salivary responses, terms like "extinction" and "spontaneous recovery" have a fairly clear-cut meaning; but one may well argue that this meaning alters when these terms are used to discuss what happens to consonant trigrams. Our own program was an attempt to fill this gap.

What follows is an account of our work during the last 8 years and a discussion of the relevant literature, directed at two problems: (1) What, if anything, do animals forget? (2) Why do they forget? In particular, how do the findings bear upon interference, stimulus-change, and decay as theories of forgetting?

I. What—If Anything—Is Forgotten?

Do animals forget? Until fairly recently, the literature provided only scattered accounts. Most workers in the area have been so impressed by the fact that learned patterns persist, that few have asked whether they persist in full. Thus Skinner (1950) is often cited as having shown that pigeons can maintain an operant response over several years; only rarely is it mentioned that retention was far from perfect, with losses up to 75%. The notion that habits are essentially permanent—if well protected from interference and stimulus-change—was certainly widely held (e.g., Kimble, 1961, p. 281), but has never really been supported by solid empirical evidence. The few early studies on animal forgetting often lacked necessary controls for postcriterial drops, warm-up decrement, drive-state, and distribution of practice, and were thus quite inconclusive. Our first task then was clear enough. We had to perform a series of experiments (with appropriate controls for the various possible artifacts) to answer a simple question: What, if anything, do animals forget?

What is meant by forgetting? The term has so many connotations from the common language that some self-conscious definition-hunting may be quite in order. To us, forgetting refers to some kind of performance decrement that can be attributed to the effect of a retention interval. In essence, we refer to the difference between test performance immediately after training and test performance after a longer retention interval. Many other questions arise immediately that bear upon the explanation of forgetting but that should be kept quite separate from its definition. Is forgetting permanent or is it transitory, or is it perhaps sometimes one and sometimes the other? Does the effect occur during storage or at retrieval? Is the retention loss due to "time as such," or to a change in stimulating context, or to interference? All of these are interesting and important questions the answers to which will be critical for an ultimate understanding of what forgetting is all about. But as we will use it here, the term "forgetting" refers simply to some performance decrement in a learned activity which is a function of time. It is in this sense that we ask the question: Do animals forget?

Except for one study on goldfish, all of the experiments here reported used rats as subjects; in most of them the drive was hunger and the reward was food. Some aspects of the procedure were the same for all studies that employed food-motivated rats:

1. The Ss were male, pigmented adult rats, ranging in age from 100 to 180 days at the start of each experiment. They were always run on the same deprivation schedule, deprived for about 23 hours and maintained at 85% of their adjusted bodyweight. They were placed on this schedule soon after they arrived in the laboratory and were maintained on it during training, during the retention interval, and during test.

2. The retention interval of the experimental groups was usually in the range of 30–60 days while that of the control group was always 24 hours. We chose a 24-hour control interval in preference to intervals that are shorter, since this equates across groups for possible satiation effects, activity levels, and warm-up decrements.

3. Insofar as at all possible, the overall stimulus situation remained the same for training, retention interval, and test. The Ss lived in the same cages throughout the experiment, remained in the same room and were cleaned and handled at the same time of day. The number of animals in the experimental room stayed roughly constant and considerable care was taken to ensure equality of visual, auditory, and olfactory cues at every stage of an experiment.

4. The method employed during the test trials was usually relearning, with a procedure identical to that used in acquisition. In this manner both "recall" (performance on the first test trial) and "relearning" (performance on the trials thereafter) could be assessed.

5. In a few of the studies a control group was run to guard against possible confounding effects due to differences in age, length of stay in the laboratory, and length of time on the deprivation schedule. This group was brought into the laboratory and placed on the drive schedule at the same time as was the experimental group. The normal control group and the experimental group would always receive their original training at the same time, the control group being tested one day after training and the experimental group m days thereafter. The age-drive control, however, would receive its training m days later than the other two groups; consequently, its test trials would occur on the same days as those for the experimental groups. On test trials both groups would therefore be equated for age, length of time on the deprivation schedule, and

length of time in the laboratory. As it turned out, the results for the age-drive control groups were virtually identical to those obtained for the corresponding normal controls. In consequence, we eventually abandoned the age-drive groups as unnecessary.

We shall first turn to those of our studies which show that rats indeed forget in many situations, and second to those which show that they do not forget under all conditions.

A. CONDITIONS UNDER WHICH FORGETTING OCCURS

1. Forgetting of "Excitatory" Response Tendencies—The Runway

We have repeatedly found that rats forget simple instrumental responses like running in an alley. The results of a representative experiment are presented in the left-hand panel of Fig. 1 (Gleitman & Steinman, 1963). Twenty Ss were

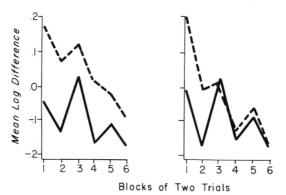

Blocks of Two Trials

Fig. 1. The effect of retention interval and prior extinction on runway performance (——— 1 day, - - - 64 days). Results are presented as difference scores in log seconds between performance on test and on the last four trials of acquisition. The panel on the left gives results for Ss without prior interference, that on the right for Ss who were extinguished prior to reconditioning. (From Gleitman & Steinman, 1963.) [Copyright (1963) by the American Psychological Association, and reproduced by permission.]

given 20 trials (4 trials per day for 5 consecutive days) on a 6-foot runway, and tested either 1 day or 64 days after training. The control group was placed on the deprivation schedule on the same day as the experimental group, but was trained 64 days later, thus equalizing age and drive state on the test trials. Performance was assessed by measuring the total time required to traverse the runway. The results for 12 test trials (4 trials per day for 3 consecutive days) are presented as difference scores from the Ss' performance on the last four trials of training. Performance was evidently a function of retention interval, an effect present both on the first test trial ("recall") and on the 11 reinforced test trials thereafter ("relearning").

It should be noted that this general result has been obtained before, but typically in situations which did not rule out various artifacts. For example, Gagné (1941) found increasing retention losses of runway performance in rats with intervals of 3, 7, 14, and 28 days. His results are open to question since his *S*s were given *ad lib.* access to food during the retention interval so that any subsequent performance decrement might well have been caused by a drop in drive level. Other examples are provided by studies of Finger (1942) and by Mote and Finger (1943) which showed retention decrements after a 24-hour interval. Their results are hard to evaluate because there were no separate control groups with a shorter interval.

2. Forgetting of "Inhibitory" Response Tendencies—Fixed-Interval Performance

In the runway, forgetting is manifested as a decrease in response strength (i.e., decrease in speed). The question arose: Would forgetting be manifested as an *increase* in response strength if what was learned originally involved the suppression of a response? Apart from its inherent interest, the answer to this question is relevant to the suggestion that decrements in runway performance after an interval of several weeks reflect a change in motivation rather than a memorial effect (e.g., Ehrenfreund & Allen, 1964). To get at this point we trained 17 rats in a modified Skinner box on a fixed-interval schedule (FI-1) for 3 days with 50 reinforcements on each day (Gleitman & Bernheim, 1963). Performance was assessed by the ratio of responses made during the first 30 seconds of the interval to the number made during the entire interval (first-half ratio). After the third day of training on the schedule, the *S*s were matched on the basis of the ratio they achieved on that day, and tested either 24 hours or 24 days following training, using the identical procedure operative during training. Figure 2 indicates the results of the two groups on the last training day and on the test day. As the figure indicates, the ratio increased with longer retention interval ($F = 17.0$, $df = 1/15$, $p < .01$): The animals forgot *not* to press immediately after reinforcement. There are many situations in which animals learn both what to do and what not to do; to the extent that this holds for learning, it evidently does for forgetting as well.

3. Forgetting of the Stimulus Situation

a. *Generalization Decrement as a Function of Retention Interval.* The preceding studies showed that rats forget response characteristics of an instrumental situation; depending upon the situation, they would forget both what to do and what not to do. It was natural enough to inquire about the forgetting of the stimulus.

To test whether rats forget incidental stimulus features of the training situation one may confront them with an altered stimulus arrangement sometime after original training. This line of attack was first employed by

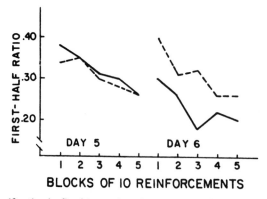

Fig. 2. First-half ratios in fixed-interval performance as a function of retention interval (——— 24 hour group, - - - 24 day group). First-half ratio = number of responses during first 30 seconds of the interval divided by number of responses during the entire interval. (From Gleitman & Bernheim, 1963.) [Copyright (1963) by the American Psychological Association, and reproduced by permission.]

Perkins and Weyant (1958) who ran rats in either a black or a white runway and then tested them on either the same runway or the one differing in color, after either 1 minute or 1 week since acquisition. They found that disruption of performance on the new runway was considerably greater immediately after learning than 1 week thereafter, an effect they attributed to forgetting of runway color. Some aspects of their procedure make firm conclusions difficult: Training was by intermittent reinforcement and test trials were unreinforced. It seemed quite possible that these variables might interact with the retention of performance in a new runway, all the more so since there is evidence that resistance to extinction increases and that the partial reinforcement effect decreases with increasing time intervals (e.g., Aiken & Gibson, 1965; see also p. 15 of this chapter).

Steinman (1967) used continuous reinforcement during both acquisition and retention tests. Eighty rats were employed in a 2 x 2 design with retention interval (1 vs. 66 days) and stimulus conditions (test trials in the same runway used during training vs. runway of a different color) as the two factors of the experiment. Both original training and retention tests were conducted at the rate of one trial per day for 12 trials. The primary result was a striking interaction between retention interval and the effect of stimulus change ($F = 11.43$, $df = 1/72, p < .01$). When compared to Ss tested 1 day after training, Ss tested after 66 days showed the expected decrease in running speed if tested on the same runway but an *increase* if tested on a runway of a different color. These results are summarized in Fig. 3; they obviously confirm the findings of Perkins and Weyant (1958). It appears that the rats have forgotten the color of the runway on which they were run initially. Having done so, they no longer

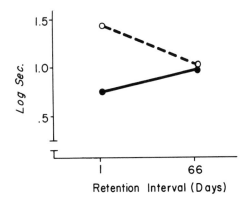

Fig. 3. Response to changed stimulus conditions as a function of retention interval, based on performance on the first test trial (●————● unchanged runway, ○- - -○ changed runway). (From Steinman, 1967.) [Copyright (1963) by the American Psychological Association, and reproduced by permission.]

respond to a new color as novel, for to do this they must first recognize that it is different from the old.

Similar effects have been obtained with rats given avoidance training (Desiderato, Butler, & Meyer, 1966; McAllister & McAllister, 1963). Equivalent results have been found for pigeons in operant situations in the form of a flattening of the generalization gradient (Honig, 1969; Thomas & Burr, 1969; Thomas & Lopez, 1962). The failure of Thomas, Ost, and Thomas (1960) to find such an effect was probably due to their use of a warm-up period prior to generalization testing. As Thomas and Burr (1969) have shown, this procedure abolishes the effects of the retention interval on the generalization gradient. (We might add that these last four studies employed intermittent reinforcement during training and unreinforced trials during the retention test, procedures we have already criticized.)

  b. *The Method of Testing Retention.* In the field of verbal learning, it is a virtual truism that the degree of forgetting obtained in a given situation is, in part, a function of the manner in which retention is tested. Thus, Postman and Rau (1957) measured the retention of serial lists of words or nonsense syllables by the methods of recognition, free recall, and relearning and found large and stable differences as a function of the method used (e.g., recognition invariably produced the highest retention scores). Does something analogous hold for our situation? As we have seen, in Steinman's study (1967) there was no hint of a difference after a 66-day interval between the animals tested on the same runway and those tested on another. Steinman showed that given another method of testing, some retention was demonstrated after all.

  The new retention test began after the twenty-fourth trial (12 during training,

Table I

Experimental Design and Results of Steinman's Study[a]

Group	Retention interval between Trials 12 and 13[b]	First retention test (Trials 13–24)		Second retention test (Trials 25–36)	
		Runway color	Mean total time on Trial 13 (in log seconds)	Runway color	Mean total time on Trial 25 (in log seconds)
AA-1	24 hours	A		A	.58
			.73		
AB-1	24 hours	A		B	1.52
BB-1	24 hours	B		B	.59
			1.45		
BA-1	24 hours	B		A	.77
AA-66	66 days	A		A	.64
			1.10		
AB-66	66 days	A		B	1.35
BB-66	66 days	B		B	.54
			1.19		
BA-66	66 days	B		A	.72

[a] A = runway color during acquisition, B = the opposite color.

[b] The retention interval between all other trials was always 24 hours.

12 during the first retention test). Now half of the *S*s received another 12 trials on the same runway employed during Trials 13–24; the other half were tested on the runway of the opposite color from that which was used during those trials. (Except for the retention interval between Trials 12 and 13, all intertrial intervals were 24 hours.) The experimental design is summarized in Table I. The critical comparison is between groups that now suffer their first shift in runway color (e.g., black during acquisition, black on the first retention test, white on the second) and those that are shifted for the second time, and in fact are shifted back to the color employed during acquisition (e.g., black, white, and black again). If the acquisition color had been completely forgotten (in the groups with a 66-day interval between trials 12 and 13), then *S*s shifted back to the old runway should show the same degree of disruption as those *S*s that were shifted to a runway they had never experienced before.

The results for Trial 25 are presented in the last column of Table I. As indicated in the table, a change in color produced a marked decrement if the color was totally new (e.g., black on Trials 1–24, white on Trial 25), but a milder one if the change was to a color previously experienced (e.g., white on Trials 1–12, black on Trials 13–24, and white again on Trial 25). Evidently the "new" runway was recognized as the old one from before. The point of interest for us is that the return to the old color produced a smaller decrement regardless of the length of the retention interval between Trials 12 and 13. There was a significant interaction between the effect of present change and prior change both for those Ss for whom Trials 12 and 13 were separated by 24 hours ($F = 8.42$, $df = 1/32$, $p < .01$) and for those Ss for whom these trials were separated by 66 days ($F = 5.25$, $df = 1/32$, $p < .05$). This fact suggests that some information about the original runway color *was* retained after all.

Why does the first retention test show so much more forgetting than the second? Is recognizing that something has changed more difficult than recognizing that it is still the same? We do not know. Whatever the explanation, it is interesting to note that for rats no less than for college sophomores, amount retained is in part a function of the method whereby retention is measured.

4. Forgetting of Reward Magnitude

The design of the next experiment utilized the well-known "depression effect" first reported by Crespi (1942). Ss who receive a less preferred reward typically perform better if rewarded thus from the outset; if first rewarded more handsomely, their later performance under the less preferred reward is usually inferior. This depression effect clearly implies that in some sense Ss "remember" the reward they had received initially. Whether the actual effect of incentive magnitude is thought to be due to an anticipatory goal response (Spence, 1956) or due to an expectation (Jones, 1952) is quite irrelevant to the problem of its retention: Be it r_g or be it idea, to be effective it must be retained over time. The present concern is whether it may also be forgotten.

We performed a runway experiment (Gleitman & Steinman, 1964) which employed six groups of about 12 rats each. All Ss received 23 trials on the runway, administered on consecutive days. Groups L-1 and H-1 were the standard ones usually employed to study the depression effect. Group L-1 was run at a low reward magnitude (two pellets) on all trials; Group H-1 was run at a high reward level (20 pellets) during Trials 1-12 but from Trial 13 on it received the same low reward as did Group L-1. The performance of these two groups is shown in the top panel of Fig. 4. Comparing the results on Trials 14–23 we note the usual depression effect: Group H-1 ran more slowly than Group L-1 ($F = 4.70$, $df = 1/22$, $p < 0.5$).

The procedure for Groups H-68 and L-68 was identical in all respects but one to that followed for Groups H-1 and L-1, respectively: a retention interval of 68 days was interposed between Trials 12 and 13. The results are shown in the middle panel of Fig. 4. As the figure shows, the depression effect failed to develop. The effect of the interposed retention interval was supported by an analysis of variance on the postshift performance of Groups H-1, L-1, H-68, and

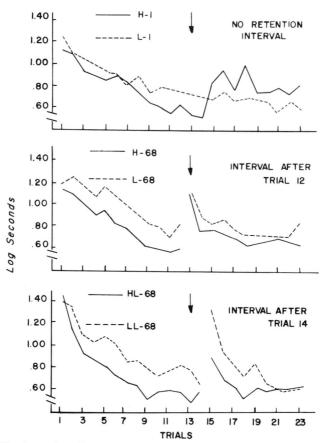

Fig. 4. The depression effect as a function of retention interval before and after shift of reward magnitude (—— shifted, - - - - - - not shifted). Arrow indicates trial on which high-reward animals were shifted to low reward. (From Gleitman & Steinman, 1964.) [Copyright (1963) by the American Psychological Association, and reproduced by permission.]

L-68 which yielded a significant interaction between original reward magnitude and retention interval ($F = 4.16, df = 1/44, p = < .05$).

The results for Groups H-68 and L-68 indicate that the Ss forgot the prior reward magnitude. We were interested to determine whether they would also forget their own reaction to the downward shift. Accordingly we ran two more groups, HL-68 and LL-68, treated exactly like Groups H-68 and L-68 except that the 68-day interval was imposed two days *after* the shift, that is, between Trials 14 and 15. (The placement of the interval was based on the fact that the depression effect was first manifested on Trial 15 for the two control groups, H-1 and L-1.) Their performance is shown in the bottom panel of Fig. 4; again, there is no trace of a depression effect. Comparing performance on Trials 15–23 for Groups H-1, L-1, HL-68, and LL-68 we again found a significant interaction between prior reward magnitude and retention interval ($F = 4.90$, $df = 1/42$, $p = < .05$).

The main findings of this study have been confirmed and extended by Gonzalez (1970), who found a progressive decline in the depression effect as he extended the interval between the preshift and postshift phase from 26 to 68 days. The general conclusion seems safe enough: Ss forget a previously experienced reward magnitude and for that matter, their own reaction at finding the reward diminished.

What accounts for the retention loss? An obvious hypothesis is that there is interference from reward levels experienced in the home cage. This could be tested quite directly by appropriate manipulations of home cage feedings, but such an experimental test has not been performed so far. However, a study by Spear and Spitzner (1968) provides indirect evidence which casts some doubt upon the hypothesis as stated. They found that contrast effects were larger if the preshift reward level followed normally upon the animal's run into the goal box; if the animal was merely placed in the goal box during the preshift period, the contrast effects were attenuated. This result suggests that a particular reward level is primarily contrasted to reward levels encountered in the same specific situation. If this is so, then it seems quite unlikely that reward levels experienced altogether outside of the runway situation would enter the "adaptation level" to reduce contrast effects.

The experiment just discussed assessed retention of reward magnitude by observing performance when the reward level was changed. What happens when performance is measured after a retention interval, while keeping the original reward levels unchanged? Steinman performed a runway experiment using a 2 x 2 design with reward level (2 vs. 10 45-mg pellets) and retention level (1 vs. 66 days) as the two factors (Steinman, 1967). She found the usual superiority in running speeds for Ss who obtained the greater reward magnitude, a superiority that was still present on the first test trial after a 66-day interval. It appears that while forgetting of reward magnitude is readily apparent when we test by a change in the reward conditions, it is not manifested when we test keeping the original conditions intact (similar results have been obtained by Gonzalez, 1970). The parallel to the findings on the retention of the stimulus situation (p. 6) is very tempting; there, too less retention was shown when the test condition involved a change from the original condition than when it involved a return to the original condition.

B. CONDITIONS UNDER WHICH FORGETTING DOES NOT OCCUR

The preceding studies give ample evidence that learned patterns in animals are not exempt from retention loss. Rats forget to run quickly in an alley, they forget not to press immediately after reinforcement in a fixed-interval schedule, they forget the color of the runway on which they were run, and they forget the amount of reward to which they had previously become accustomed. The variety of things that rats can forget is rather impressive. Our next question is, do they forget under *all* conditions?

1. Retention of Choice Situation

All of the studies reported thus far have one thing in common: The learned response is of the "go, no-go" variety, and learning—and forgetting—are assessed by changes in latency or rate. For such situations, forgetting was always found in our studies. What happens when the task involves a choice?

a. Spatial Discrimination. Our first study (Gleitman & Jung, 1963) employed 20 rats trained in a modified Skinner box with two pigeon keys on one wall and a lever on the other. Each session consisted of 64 discrete trials with a 15-second intertrial interval. To start any trial, the subject first had to press the lever in order to light up the keys. The task then was to press the correct key 10 times (not necessarily in succession). The trial was over and the reward was presented when 10 correct key-presses had been accumulated. This procedure allowed us to utilize a more sensitive measure of discrimination than a dichotomous choice score—the number of false responses per trial.

Four groups of five animals each were used. All of the animals first learned to choose one key rather than the other. Training continued until they reached a criterion of 58 or more errorless trials in any one daily session. Two of the groups (N-1 and N-44) were "nonreversal" groups: They received no further training but were tested either one day or 44 days following the criterion session. The retention test consisted of one relearning session, in which the procedure was identical to that used during training. The remaining two "reversal" groups (R-1 and R-44) were trained to reverse the spatial discrimination which they had first mastered. They were trained until they reached the same criterion as before; retention of the second discrimination was tested either 1 day or 44 days after they reached the required standard.

Figure 5 presents mean error scores (over two blocks of 32 trials each) for the last acquisition session and the test session. As the figure shows, there was no retention loss unless prior interference was provided. An analysis of variance was performed on the difference scores for errors on test and final training sessions, yielding a significant interaction between retention period and presence of prior interference ($F = 11.01$, $df = 1/16$, $p < .01$). There was rapid relearning, but a considerable effect was nevertheless still evident in the second half of the test session. The interaction was again significant ($F = 4.98$, $df = 1/16$, $p < .05$).

For the time being we want to emphasize the results for the two groups that did *not* suffer prior interference. It appears that for a simple spatial discrimination there is little or no forgetting unless the Ss have previously

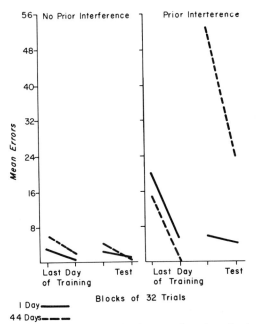

Fig. 5. Performance in a spatial discrimination as a function of retention interval and proactive interference. (From Gleitman & Jung, 1963.) [Copyright (1963) by the American Association for the Advancement of Science.]

learned the opposite response. The fact that retention loss was enormous given proactive inhibition (PI) is, of course, highly relevant to one version of the interference theory of forgetting and will be considered later in that context.

b. Visual Discrimination. Would the results be the same for a visual discrimination? To answer this question we ran a rather similar study, this time requiring the subjects to choose between horizontal and vertical stripes (Maier & Gleitman, 1967).

The experiment employed 48 rats run in standard test chambers equipped with response keys upon which the visual patterns could be projected from behind. Following pretraining, both patterns were presented simultaneously, their position determined by a Gellerman order. All Ss first learned one discrimination in daily sessions of 40 trials per day until they reached a criterion of 32 or more correct trials in a session. They were then matched by "days to criterion" and assigned to reversal (R) and nonreversal (NR) groups. Half of the NR Ss, again matched by days to criterion, were tested one day after the criterion day (NR-1), and half were tested 32 days later (NR-32). All Ss in the R groups were trained to reverse their original discrimination until they reached the same criterion of 32 out of 40 trials a day. Again matched by days to criterion, half of the subjects in the R groups were tested 1 day later (R-1) and

half were tested 32 days later (R-32). On the test day, the Ss were confronted by
the last discrimination to which they had been exposed. The test day consisted
of 40 further acquisition trials of this discrimination.

Figure 6 presents mean errors across 10-trial blocks on the last day of
acquisition and on the test day, for the last discrimination learned. As the figure
indicates, the results are in close accord with those obtained for right-left

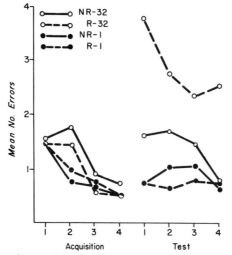

Fig. 6. Performance in a pattern discrimination as a function of retention interval and
proactive interference. Mean number of errors in 10-trial blocks for reversal and nonreversal
groups on last day of acquisition and on test. (From Maier & Gleitman, 1967.)

discrimination: There was little retention loss without prior interference.
Comparing the test performance of individual groups we find that Group R-32
made significantly more errors than any of the other three groups ($p < .001$);
none of the other comparisons yielded significant results. Again, our present
concern is with the effect of time interval in cases where specific proactive
interference is not provided. Under those conditions forgetting is either absent,
or so negligible as to be lost in the statistical noise.

    c. Some Problems of Interpretation. It appears that in simple choice
situations forgetting is hard to find. Does that mean that in these situations there
is *no* forgetting? The evidence suggests that this would be overstating the case.
True enough, both we and several other investigators have failed to obtain much
of a retention loss for simple choice behavior after intervals ranging from 5 to 44
days. (These results were usually obtained in the course of running control
groups for studies of proactive or retroactive inhibition.) But it is worth noting
that in all of these studies such slight differences as were obtained were in the
direction of forgetting (Chiszar & Spear, 1968a; Gleitman & Jung, 1963; Kehoe,

1963; Maier & Gleitman, 1967) and sometimes significantly so (Crowder, 1967). A more recent study showed a significant increase in errors on a T-maze after a 14-day interval (Hill, Cotton, Spear, & Duncan, 1969); interestingly enough, the forgetting effect was more pronounced when the measure was running speed in the T-maze arms rather than number of errors.

These facts suggest that the comparative stability of simple choice situations over time may be an artifact of the choice measure itself which may be less sensitive to variations in the strength of the individual response tendencies than are the latency measures of the two responses when taken separately. If this is so, slight changes in individual response strength will be picked up by go, no-go measures but not by choice.

The preceding comments apply to simple choice situations, where what is to be learned involves only one choice. The pattern of results is rather different when the choice situation is more complex and requires that several choices have to be learned and kept memorially distinct from each other. Here forgetting is often very pronounced. This is the case for the retention of multiple-unit mazes, which has been shown to decline rather sharply by Tsai (1924), Magdsick (1936), and Bunch (1941) over intervals up to 120 days. This may well relate to the fact that, as already mentioned, sizable retention decrements have been obtained when proactive interference was provided (e.g., Gleitman & Jung, 1963), though this result is by no means found by all investigators (e.g., Kehoe, 1963).

2. *Forgetting as Assessed by Resistance to Extinction*

In all of the preceding studies, retention was tested by relearning. It is natural enough to choose extinction as another kind of test. Our study used a simple rat test chamber with a retractable lever (Alexander, 1966). Twelve rats were run on discrete trials with a fixed intertrial interval of 1 minute. At the beginning of each trial the lever became available; as soon as it was pressed it was withdrawn and the (hungry) S received a 45-mg food pellet. All Ss were run for three sessions with 30 trials per session and with latency the primary measure. Subsequently they were divided into two groups, matched by terminal performance, and extinguished either 1 day or 52 days after the last day of acquisition. During extinction, a trial was terminated by a lever press or if S failed to respond within 90 seconds ("balk" trial). Ss were run for 2 days under extinction. Each extinction session lasted 100 trials or 110 minutes, whichever was reached first. Figure 7 shows the average number of trials required by animals in the two groups to reach successively more stringent criteria of extinction. It appears that resistance to extinction *increases* with increasing retention interval.

Essentially the same result had been obtained by earlier investigators (Brady, 1951; Youtz, 1938) but little was made of it, perhaps because of the then widely held belief that resistance to extinction is a valid means of assessing the strength of a habit. More recent workers have considered the phenomenon in the context of retention processes. Aiken and Gibson (1965) found greater resistance to extinction in animals extinguished 21 days after training under continuous reinforcement. (Animals trained on an intermittent reinforcement schedule gave, if anything, the opposite result.) They interpret their results in

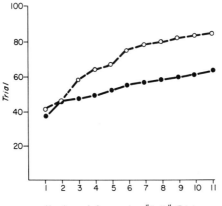

Fig. 7. Mean number of trials ($-\!\circ\!-$ delayed, $-\!\bullet\!-$ immediate) to reach successively more stringent extinction criteria (1–11 successive "balk" trials in which S failed to respond within 90 seconds). (From Alexander, 1966.)

the light of the discrimination hypothesis of the partial reinforcement effect (PRE), proposing that the continuously reinforced animals forget various aspects of the acquisition situation and so have greater difficulty in discriminating acquisition and extinction. Their interpretation has been challenged by Spear and Spitzner (1967) who found increased resistance to extinction after an 8-day interval in a group of rats originally trained under continuous reinforcement, but obtained no such effect in another group that was first given a series of nonrewarded trials before rewarded acquisition training ever started. Spear and Spitzner argue that this effect of initial unrewarded trials is difficult to account for in terms of the discrimination hypothesis: Why should initial nonreward make it more difficult to discriminate extinction from acquisition, given that both groups had the same number of continuously reinforced trials as their last experience on the apparatus just before extinction, and thus the same opportunity to discriminate acquisition from extinction?

What other interpretation can we offer for the apparent fact that resistance to extinction increases with retention interval? We think there may be a clue in the fact that we did *not* obtain this effect in a recent pilot study in which the reward magnitude during acquisition was rather small. [Similar results were obtained by

Gonzalez (1970) who also reached very similar conclusions.] Consider the evidence on the effect of reward magnitude on extinction, under the usual condition where the interval between acquisition and extinction is short. Generally, rats have been found to extinguish faster the larger the original reward magnitude (e.g., Wagner, 1961). In a recent paper Gonzalez and Bitterman (1969) argue that this phenomenon can be understood as a contrast effect analogous to that which Crespi (1942) called "depression." It is known that the depression effect increases as a function of the discrepancy between the pre- and postshift magnitudes (e.g., Gonzalez, Gleitman, & Bitterman, 1962). If extinction represents a downward shift (from something to nothing), this of course becomes more pronounced the greater the original reward magnitude. Given this assumption, the effects of reward magnitude upon extinction follow directly. So indeed do our present results.

We have already shown that the Crespi depression effect is abolished by a 2-month retention interval (Gleitman & Steinman, 1964). If it is true that extinction is deepened by a depression effect of just this kind, the conclusion is obvious: A retention interval must increase resistance to extinction. The fact that we did not obtain this result with low reward magnitudes fits very nicely: With low reward magnitudes there is no depression even when the acquisition–extinction interval is very short. Thus there is no effect that can dissipate over time.

3. Retention of the Conditioned Emotional Reaction

All of the studies discussed thus far have employed appetitive motivation, with hungry rats rewarded by food. What happens when the motivation is aversive?

We have run one study directed at this issue (Gleitman & Holmes, 1967), involving the retention of a conditioned emotional reaction (CER). In a widely cited experiment (Hoffman, Fleshler, & Jensen, 1963) pigeons showed no forgetting of a CER after an interval of 2½ years. We believed that this finding might well be a procedural artifact since Hoffman et al. used highly overtrained subjects who served as their own controls and received several days of reinstatement on the original baseline prior to the retention test. We tried to control for these factors by testing for the retention of a very incompletely learned CER, and by the use of separate groups for the two retention intervals.

The Ss were 30 rats, trained on a fairly standard CER procedure, adapted from Kamin (1965). The Ss were hungry and reinforced by food; the apparatus was a standard test chamber. After five 2-hour sessions on a VI 2.5 schedule, the Ss received four CS–UCS pairings, presented during another 2-hour session that was in all other respects identical to the five preceding. The CS was white noise at 80 dB of 3-minute duration; the UCS was a 0.5-second, 1-mA foot shock, presented immediately upon termination of CS. Responses were recorded during the 3 minutes while the CS was presented (C) and during the 3 minutes just prior to its presentation (P). The CER was assessed by the suppression ratio C/(P+C). Half of the Ss were tested 1 day following the training session (Group A); the other half were tested 90 days later (Group B). The procedure on the test session was in all respects identical to that employed on the training session.

The results are presented in Fig. 8. Trials 1–4 indicate suppression during the

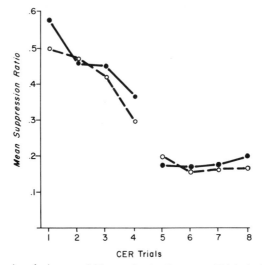

Fig. 8. Suppression during acquisition and retention tests. Trials 1–4 are the training session, Trials 5–8 the test session. Group A (●—●) was tested 1 day and Group B (○- - ○) 90 days after training. (From Gleitman & Holmes, 1967.)

training session, Trials 5–8, for the session on which retention was tested. It is obvious that suppression was readily acquired, but clearly not overlearned. It is apparent also that there was no sign of a retention decrement $(F < 1)$.

It would seem that Hoffman, Fleshler, and Jensen's finding is genuine enough and cannot be ascribed to artifacts of their procedure. What then can account for the perfect retention of the CER, given the fact that substantial forgetting occurs in many other learning situations? One might propose the rather sweeping hypothesis that any learning based upon very intense or aversive reinforcing stimuli becomes thereby immune to forgetting. This assertion is certainly too broad. At the least it fails to hold for aversively controlled instrumental conditioning since Deutsch and Leibowitz (1966) found considerable forgetting of a Y-maze habit reinforced by escape from shock. A more limited (and more plausible) hypothesis would be that enhanced retention is a general characteristic of classical aversive conditioning. On this point the evidence is more ambiguous.

Several studies have obtained virtually perfect retention of classical aversive conditioning in adult rats after intervals of up to 50 days. Thus Campbell and Campbell (1962) found no forgetting of a learned fear as measured by conditioned suppression; Kirby (1963) found no appreciable decrement in an avoidance response on a runway. Campbell and Campbell (1962) subjected rats to repeated inescapable shocks in an experimental compartment and later tested for the retention of learned fear by allowing the animal a choice between a "safe" compartment and the one in which it was shocked; in adult rats retention was essentially complete.

Other investigators have obtained rather different results. Gleitman and Rozin (see p. 37 of this chapter) trained goldfish on a signaled avoidance in a shuttle box and obtained massive retention decrements after intervals of 4 or 8 weeks, a result that has stood up after several replications. Spear (Chapter 2) presents evidence for some forgetting of active avoidance in adult rats, though the loss was slight compared to that found in infant animals. McAllister and McAllister (1968) used a procedure quite similar to that employed by Campbell and Campbell: Their Ss first received inescapable shock in one compartment and later had the opportunity of jumping a hurdle to escape from this compartment. Unlike Campbell and Campbell, they found a substantial retention loss, and this with an interval of only 47 hours. It is worth noting that McAllister and McAllister gave their animals two shock-free familiarizations in the apparatus before fear-conditioning began; this may well have provided the conditions for the development of proactive inhibition. (No such preexperimental exposure to the apparatus occurred in the Campbell and Campbell study.)[2]

----

[2] We do not make reference to the numerous studies of the "Kamin effect," a U-shaped function often obtained for the retention of aversively motivated behavior with peak performance at 0 and 24 hours after acquisition and the bottom at between 1 and 4 hours (Kamin, 1957; for a recent review, see Brush, 1971). Until recently most investigators of the Kamin effect assumed that the phenomenon is caused by some kind of motivational change (e.g., "incubation of fear," Denny & Ditchman, 1962; "parasympathetic overshoot," Brush, Myer, & Palmer, 1963) rather than by memorial events. If this is true then experiments on the retention of aversively based learning using intervals between 0 and 24 hours can be safely ignored by the investigator primarily interested in memorial processes. However, Klein and Spear (1970) have recently questioned the motivational basis of the Kamin effect, arguing that it should be reinterpreted as an instance of retrieval failure. They employed a negative transfer paradigm, training Ss on an active avoidance followed at various intervals by passive avoidance, or the other way around. They found massive negative transfer either immediately after original training or 24 hours later with no deleterious effect during the intermediate periods. Since the results were in the opposite direction of that usually obtained when the retention test involves the same avoidance response acquired originally, the authors conclude that what declines during the intermediate intervals is the memorial availability of the first-learned habit: Failure to retrieve the first habit will then prevent negative transfer in acquiring the second.

Klein and Spear's analysis is very ingenious but it is not yet conclusive for it may be

II. The Why of Forgetting

The studies we have presented thus far make it clear that rats will forget under many conditions. The next step is a search for the explanations of forgetting. Several theories have been offered: (1) interference, be it proactive or retroactive, (2) change of stimulus or context, and (3) decay. We will deal with each of these in turn, presenting our experiments as they bear upon each.

A. INTERFERENCE

Interference theorists are agreed that, in one fashion or another, forgetting is essentially a species of negative transfer. We forget because we keep on learning; in the apocryphal words of an ichthyologist turned university president who tried to call all his students by name: "Every time I learn another student's name, I forget the name of a fish" (Hilgard & Atkinson, 1967, p. 323). At least on the face of it, interference theory needs no principles other than those that govern learning and transfer to explain the phenomena of forgetting. This of course has been one of its major appeals and probably helps to account for its extraordinary popularity despite the fact that the evidence in its favor has been scanty thus far.

1. Retroactive Inhibition

"Classical" interference theory (McGeoch, 1932) regarded retroactive inhibition (RI) as the major villain. This version of interference theory has the advantage of coping most naturally with what is after all the primary task of any theory of forgetting: to explain the deleterious effect of the retention interval. According to this theory, the interval provides the opportunity for the (extra-experimental) acquisition of competing responses which make their weight felt on the test. The longer the interval, the more will be the opportunities for such competing learning to occur.

The arguments against this version of interference theory were brilliantly stated by Underwood (1957). They come from the fact that RI increases with

---

possible to accommodate their results within a motivational theory of the Kamin effect. Considerations such as Spence's (1958) lead one to the prediction that the acquisition of a new response will be retarded by increases in the drive level if competing responses are at a high level of learning. Evidence in line with this prediction has been obtained in studies of verbal learning (e.g., Spence, Farber, & McFann, 1956) and in several reversal learning experiments using animal subjects (Gossette & Hood, 1967; Kendler & Lachman, 1958). If one assumes that fear is greater at 0 and 24 hours after original avoidance learning and at a diminished level during the intermediate intervals, then the U-shaped negative transfer function obtained by Klein and Spear follows, given that their retention test was a reversal of the first-learned task so that competing responses were presumably at high strength.

increasing similarity between tasks but that massive retention losses are often seen, nevertheless, for materials which seem very different from whatever competing material one might reasonably assume the S to be exposed to during the retention interval. Chess players lose their sharpness if they have not played for a while; what could they possibly have learned during the interval that is similar enough to the chess situation to make them forget the nuances of a queen's pawn opening? This argument applies even to nonsense syllables. In Underwood's words:

> . . . it seems to me an incredible stretch of an interference hypothesis to hold that this 75 percent forgetting was caused by something which the subjects learned outside the laboratory during the 24-hour interval. Even if we agree with some educators that much of what we teach our students in college is nonsense, it does not seem to be the kind of learning that would interfere with nonsense syllables . . . (Underwood, 1957, pp. 50–51).

This analysis may be quite appropriate for the forgetting of various verbal materials but that does not necessarily mean that RI should be ignored in the kinds of situation we have dealt with in animal learning. Consider the runway situation. During acquisition, the rat learns a variety of responses within the runway all of which serve to increase running speed: orienting in the start box, leaping out as soon as the door is opened, and so on. It is perfectly possible that the rat acquires some further responses in the period just before it is placed into the runway that also increase its running speed. For example, the rat learns to approach the experimenter who is about to take it out of the cage and into the alley, a response that may well be preparatory to running the alley in the sense of improving actual performance. During the 2-month retention period, such preparatory responses might extinguish: while the experimenter still approaches the cage and sometimes handles the animal (for feeding, weighing, maintenance), he certainly does not place the animal into the runway. One could then argue that the 2-month retention interval is a 60-trial extinction session for a whole set of preparatory responses. Once these are extinguished, the runway responses (which thus conceived are a later part of a response chain) necessarily collapse. [A similar suggestion, couched in terms of an r_g-s_g mechanism which affects incentive motivation, was offered by Ehrenfreund and Allen, (1964).]

We have performed an experiment directed at this possibility. The rationale of the study is simple. If interfering responses (e.g., extinction of preparatory patterns) are acquired during the retention interval, their deleterious effect should be less if the animals spent the retention interval in a *different* environment from that in which they were originally trained and later tested. By this procedure the potentially competing responses are conditioned to stimuli other than those to which the critical, original responses are connected; hence these new responses never can fully realize their interfering potential. In consequence, forgetting should be less.

Thirty-four male rats were run in this study, on the usual food-deprivation schedule. All Ss first received 12 trials (1 trial/day) on the runway. During training their living cages were in a room directly adjacent to the room in which the runway was placed. These cages were standard slide-in cages, 7 inches x 10 inches x 7 inches, with a wire-mesh floor and no top, hanging from a rack that held 60 such cages (Wahmann Mfg. Co., catalogue number LC 75-A). After the twelfth trial, the Ss were assigned to two equal groups, matched by terminal performance. The subjects in Group I spent the retention interval in the same cages they had been in all along (which remained in their old location throughout). The subjects in Group II were moved to new cages, located in another room. These new cages were larger, 9 inches x 15 inches x 9 inches (Wahmann catalogue number LC-28), and had floors filled with animal bedding. The new room was rather different from the old: It was more brightly lit, noisier, etc. The Ss in Group II were moved back to their old locations (both room and cages) 3 days prior to the retention test. The test was administered 60 days after the last acquisition trial. During the retention test, all conditions were exactly as they had been during acquisition. Figure 9 presents the results for the first retention test of both groups. There was no trace of a difference: Both groups forgot, and to a virtually identical degree ($t < 1$).

It appears that the different environments during training and retention interval had no effect on the degree of forgetting. While one can argue that the differences were not really large enough or were not along the critical stimulus dimensions, we still feel that this result renders an RI interpretation rather implausible.

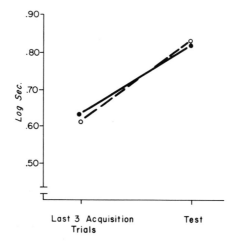

Fig. 9. Retention of runway performance in rats maintained under identical (o - - o) or changed (●—●) living quarters during the retention interval (mean total times).

2. The Role of PI in Ordinary Forgetting

Given the limitations of RI as an explanation of forgetting, interference theorists cast about for an alternative. Underwood's demonstration (Underwood, 1957) that forgetting of the last list was an increasing function of the number of other lists the *S* had learned previous to the last, focused attention upon proactive inhibition (PI) which was now held to be the major cause of retention loss. According to this view, "ordinary" forgetting (where no explicit source of interference is introduced by the experimenter) is caused by PI from extra-experimental sources of interference. These are previous habits that the *S* brings to the task which are inappropriate to it and are therefore unlearned (extinguished) during the course of original learning. They ultimately reappear due to spontaneous recovery, interfere with the correct response at the time of test, and are thus responsible for the retention decrement. In the verbal learning area, this position had originally been bolstered by studies on the temporal course of PI which showed that the deleterious effect of the first list increases with increasing time interval between original learning and test (e.g., Briggs, 1954; Underwood, 1948).

This modern version of interference theory[3] no less than its classical (RI) counterpart, must appeal to extra-experimental sources of interference to account for forgetting. Attempts to provide relevant evidence have taken different lines of experimental attack. One approach has been to manipulate the extra-experimental environment, typically during the retention interval, and thus is more relevant to classical interference theory with its emphasis upon RI. (It is obviously very difficult to perform comparable studies directed at the reduction of extra-experimental PI, since one might be called upon to manipulate the entire preexperimental life of the subject.) The usual aim has been to minimize the development of incompatible habits. Representative experiments on human subjects include the widely cited studies showing less forgetting after intervals of

---

[3] Since our own work began when "modern" interference theory was in fact still modern, and since it was so heavily influenced by those conceptions, we have kept the adjective even though by now it is really a misnomer. In fact, few students of verbal learning presently hold to the version of interference theory in the form originally proposed by Underwood (1957) and Underwood and Postman (1960). There have been failures in finding rates of forgetting that correspond to the similarity between the materials learned in the laboratory and those already in the subject's linguistic repertory (e.g., Ekstrand & Underwood, 1965) and difficulties in obtaining evidence for absolute recovery (e.g., Slamecka, 1966). In addition, several recent lines of research have cast serious doubt upon the two-factor theory of RI that in many ways had served as the critical underpinning of interference theory and historically was the start of the whole enterprise (Melton & Irwin, 1940). To give only one example, Houston (1967) showed that PI can be obtained with the method of modified free recall (MMFR), a fact inconsistent with the general scheme since PI was in theory caused by a factor (competition) which should not show up in MMFR. Evidence of this kind necessarily led to a reconsideration of theoretical premises which is still in progress (e.g., Keppel, 1968; Postman & Stark, 1969).

sleep than after equal periods of waking (e.g., Jenkins & Dallenbach, 1924; Van Ormer, 1932) and more recent work demonstrating a similar effect if the retention interval is spent under conditions of sensory deprivation (Grissom, Suedfeld, & Vernon, 1962). Related work has been performed on animal subjects. The best known study is probably that of Minami and Dallenbach (1946) who found less forgetting in cockroaches that were immobilized during the retention interval. An experiment on goldfish treated with chlorpromazine (Rensch & Dücker, 1966) gave similar results, though an earlier study on anesthetized rats did not (Russell & Hunter, 1937).

It is worth noting that whatever the results of such studies (however interesting they may be on other grounds), they cannot by themselves provide critical evidence for interference theory, whether old or new, as it has been formulated in the literature. Perhaps sleep (or immobility, or sensory deprivation) does indeed reduce retention loss but this does not prove that the effect is caused by a reduction of competing habit patterns.

A more direct approach has been to vary the nature of the material that is to be retained. According to interference theory, the laws of forgetting must be essentially equivalent to the laws that govern experimentally provided RI or PI. Given what we know about the role of similarity in those situations, it follows that the more similar the material to be retained is to the learned patterns acquired in the course of a normal lifetime, the more readily it should be forgotten. Several studies have been carried out in attempts to vary the similarity between those verbal materials learned by the subject and those most frequently encountered in the language and thus in the subject's prior history. The results have generally been quite unsuccessful (Ekstrand & Underwood, 1965; Postman, 1961; Underwood & Postman, 1960), a state of affairs appreciated most acutely by those who originated modern interference theory in the first place (e.g., Underwood & Ekstrand, 1966).

Yet another way of getting at the problem is to guess at the kind of extra-experimental interference that most probably is involved (given the assumption that such interference is relevant at all) and then somehow to vary its degree. We consider some habit for which forgetting has been shown, ask what (preexisting) reaction pattern might be a plausible source of proactive interference, and increase the strength of this reaction prior to acquisition of the habit whose retention is tested later on. If our premises are true, forgetting should now be more pronounced. This is the general approach we have taken in several of our own studies on the effect of prior interference upon forgetting of the runway and of the FI-scallop.

a. PI and Runway Performance. As previously shown, runway performance in rats declines with large retention intervals. If we accept the general assumptions of modern interference theory, we must assume that the rats enter the experimental situation with some previously established reaction patterns (perhaps based on fear or curiosity) which are incompatible with speedy running. These are presumably extinguished during the course of runway learning and recover spontaneously during the retention interval. One way to test this hypothesis is to increase the strength of these preexisting incompatible responses. If modern interference theory is correct and if we have correctly identified the preexisting source of interference, there should be greater forgetting of runway performance under these conditions since PI is known to increase with increasing degrees of prior (and incompatible) learning (Underwood, 1949).

In principle there are various means of augmenting preexisting responses incompatible with runway performance (e.g., punishment). We chose an extinction procedure (Gleitman & Steinman, 1963); a portion of this study has already been described on p. 4 of this chapter. Thirty-six rats were used in a 2 x 2 design with retention interval (1 vs. 64 days) and degree of proactive interference (presence or absence of extinction prior to final training) as the two factors. All Ss first received 20 trials (four trials per day) on the runway. For half of the Ss original training was complete at this point and test trials (all reinforced and again at four trials per day) began, either the next day or 64 days thereafter. The other half of the Ss received four unreinforced trials per day until they reached a rather stiff extinction criterion. (The S was never removed from the runway until it reached the empty goal box in which it was detained for 10 seconds. Extinction trials continued until S reached a criterion of two trials on any day in which the total time was 2 minutes or more.) Following this, Ss were retrained with the original procedure to the level of performance achieved in the last day of original training. Test trials were administered either 1 day or 64 days after retraining.

The results on the test trials are presented in Fig. 1, expressed as deviation scores from performance on the last four trials of training. The right half of the figure presents data from the animals that experienced extinction; the data from the nonextinguished groups are seen in the left half. As the figure indicates, the retention loss was *not* greater if training was preceded by extinction; if anything, the reverse was the case. Under the circumstances, the relevance of PI to forgetting on the runway becomes somewhat suspect. One could probably argue that the extinction procedure did not augment the strength of the preexisting source of interference on the assumption that this was already at an asymptotic level. While we do not find this argument particularly plausible, we certainly cannot rule it out.

b. PI and Fixed-Interval Performance. As previously described, a retention interval of 24 days leads to the disruption of the scalloping effect characteristic of fixed-interval performances: Ss forget *not* to press immediately after reinforcement. Here one can make a rather plausible case for a PI interpretation of the retention decrement. Training on the FI schedule was preceded by considerable pretraining on continuous reinforcement (CRF). The CRF pattern would be unlearned while the FI schedule was in effect and would presumably recover spontaneously over time. To settle this point, an experiment was performed using 38 rats in two groups. The Ss had no pretraining whatever, but were placed on the FI schedule from the start (Gleitman, Steinman, & Bernheim, 1965). Training continued until the groups achieved a mean first-half ratio of .33, after which they were tested either 1 day or 25 days later. The mean first-half ratios on the retention test were .26 for the 1-day group and .31 for the 25-day group, a difference that was statistically significant ($t = 2.18$, $df = 36$, $p < .05$). It appears that PI due to pretraining on CRF is not the reason why rats forget not to press in the early part of the interval on an FI schedule.

We do not wish to argue that one might not manipulate the conditions so as to produce PI of the FI-scallop by pretraining with CRF or, for that matter, to generate PI of runway performance by the administration of prior extinction sessions. As a matter of fact, there is considerable evidence for PI in certain go, no-go tasks,[4] and it would not be surprising that were we to repeat our own studies with appropriate modifications we would also find PI. But this point is not at issue. The question is not whether a particular prior experience can lead to PI of the FI-scallop or of running an alley. The question is whether such PI occurs under conditions that produce ordinary forgetting. And this does not seem to be the case.

3. Experimentally Provided PI

The preceding section suggests that there are serious obstacles in the way of a theory of forgetting based upon extra-experimental sources of proactive interference. But quite apart from that, one can raise questions about (experimentally produced) PI itself. We were concerned with whether PI could be demonstrated at the animal level and whether such investigations might reveal something about its causes.

a. Can One Demonstrate PI in Animals? A study (Gleitman & Jung, 1963) that shows PI for a right-left discrimination in rats has already been described

---

[4] To take one example, a passive avoidance produces PI of an active avoidance acquired subsequently (Spear, Chapter 2). Similarly, suppression in eating food adulterated with quinine sulfate is proactively inhibited if the animal has previously eaten unadulterated food (Chiszar & Spear, 1968b). Why the discrepancy with our findings? Spear (1967) pointed to different distributions of practice, extending over several days in our own case and a matter of minutes in the studies by Spear and his associates. Distributed practice on the two tasks should lead to reduced PI on the analogy to the verbal learning situation where precisely this has been the result (Keppel, 1964; Underwood & Ekstrand, 1966). Such an effect would also be expected if one adopts the hypothesis that PI is reduced by differential task-recency (e.g., Maier *et al.,* 1967), assuming that recency is measured from the point when training on a given task is completed.

on p. 12 of this chapter. Rats were first trained to press the right lever, then retrained to press the left one, and finally tested for the second discrimination 44 days later. The result was a massive retention loss. In contrast, no such effect appeared if the animal learned only one discrimination but was tested after the same interval. The results of this study were in conflict with those obtained by Kehoe (1963) who could find no direct evidence for PI using pigeons trained on a 5-choice discrimination. Kehoe suggested that the critical factor was her use of the noncorrection method as opposed to the correction method typically employed in studies of verbal learning (and also used by Gleitman and Jung).[5] To check on this point we conducted another experiment with an essentially similar design, but using a visual discrimination and a noncorrection procedure. This study (Maier & Gleitman, 1967) has already been described on p. 13 of this chapter. The results were virtually identical to those reported by Gleitman and Jung. Quite clearly, PI can be obtained with both correction and noncorrection procedures.

It is worth noting that in our studies the PI effect was huge and highly significant. Since both studies used a 2 x 2 design and were subjected to an analysis of variance, the interaction between prior reversal and retention interval is the critical statistic. For the Gleitman and Jung study ($N = 20$) the relevant F-ratio was 10.74 ($df = 1/16$, $p < 0.01$); for the Maier and Gleitman study ($N = 48$) the F-ratio was 14.43 ($df = 1/44, p < .001$).

  b. *What Accounts for PI?* The occurrence of PI and its increase with increasing retention intervals has generally been attributed to spontaneous recovery of the first-task responses which were assumed to be unlearned or extinguished during the acquisition of the second task. Consider this analysis as applied to the results of the two studies just described. In these studies (Gleitman & Jung, 1963; Maier & Gleitman, 1967) the PI groups were trained on one discrimination ($S_1 +$ vs. $S_2 -$) followed by its reversal ($S_1 -$ vs. $S_2 +$), with retention of the second discrimination tested 30 or more days later. There are *two* reversals in the shift from first to second discrimination: S+ becomes negative, and S− becomes positive. It would seem that the spontaneous recovery hypothesis of PI must hold that the critical reversal is the first one: that is, changing S+ to S−. Under these conditions, S must inhibit (and presumably extinguish) a previous approach tendency while learning the second task; given the possibility of extinction, there is the possibility of eventual spontaneous

---

[5] Kehoe argued that with the correction method, the correct response is learned rather quickly and gains ascendancy over the competing habits before these are completely extinguished. In contrast, the noncorrection method assures a more thorough extinction of the inappropriate response by the time the criterion has been reached. One might thus expect less PI following training by the noncorrection method since the competing responses would then have been extinguished below the point from which they can recover spontaneously.

recovery. On the other hand, it is rather implausible that changing S− to S+ could be relevant to this process for here there is no excitatory tendency to inhibit and therefore no possibility of eventual spontaneous recovery. This kind of shift should not lead to PI if the spontaneous recovery hypothesis is correct. In the usual reversal experiment, both shifts (S+ to S− and S− to S+) are carried out together and thus their effects cannot be disentangled. We performed a study designed to separate the two shifts (Maier, Allaway, & Gleitman, 1967).

Forty-eight Ss were first trained on a visual discrimination, S_1+ vs. S_2-. Subsequently, the Ss were divided into two groups, matched by number of trials to criterion on the first discrimination. Group PM was trained on a partial reversal S_3+ vs. S_1- (plus-minus shift); Group MP was trained on the other partial reversal, S_2+ vs. S_3- (minus-plus shift). The two groups were further split up into two subgroups each matched by trials to criterion on the second discrimination. Groups PM-1 and MP-1 were tested 1 day after reaching criterion; Groups PM-32 and MP-32 were tested 32 days later. The retention test consisted of one further session on the second discrimination.

The three stimuli were a triangle (T), vertical stripes (V), and horizontal stripes (H). The first discrimination was always between T and either H or V; the second always between H and V. In all other respects, the stimulus assignments were appropriately counterbalanced. More specifically (and considering only Ss who encountered H on the first discrimination) for the PM groups S_1+ was H and S_2- was T; subsequently S_1- was H and S_3+ was V; for the MP groups S_1+ was T and S_2- was H on the first discrimination, while on the second S_2+ was H and S_3- was V.

It is clear that by the previous reasoning, PI should be less when the second discrimination involves a minus-plus shift. Figure 10 indicates that this was not the case. As the figure indicates, both PM and MP groups show sizable PI with the 32-day retention interval. The mean error scores of the MP-32 and PM-32 groups are clearly larger than those obtained for the corresponding groups tested after 1 day, MP-1 and PM-1 (the two respective t-values are 3.42 and 3.28, with df values of 17 and 18 in the two cases; for both, $p < .01$). Most important, the comparison between the MP-32 and the PM-32 groups shows not the slightest indication that plus-to-minus reversal generates more PI than a shift from minus-to-plus. It might be argued that this result does not rule out the spontaneous recovery hypothesis; perhaps what was extinguished (and later recovered) was an avoidance response to the initially negative stimulus. This interpretation does not fit our data, for we found no negative transfer when the shift was from minus-to-plus.

This result is a strong argument against the view that PI is due to spontaneous recovery of inappropriate responses extinguished in the course of learning the second task, and further adds to the burden of negative evidence which this view must shoulder. As it is, this burden is quite weighty. For example, it has been

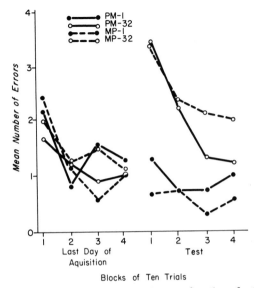

Fig. 10. Performance on a pattern discrimination as a function of retention interval and prior partial reversal. (From Maier, Allaway, & Gleitman, 1967.)

very difficult to find evidence for spontaneous recovery of the first-task responses in verbal learning situations (Koppenaal, 1963; Slamecka, 1966). In addition, there has always been the curious anomaly in the relevant time parameters: Spontaneous recovery when measured directly reaches an asymptote in a matter of hours (e.g., Ellson, 1938), whereas forgetting continues over a period of weeks. But even should all these problems be resolved, one would still have to answer a further question: What leads to spontaneous recovery in the first place? The fundamental question about forgetting concerns the role of elapsed time. To say that PI increases with time only shifts the problem elsewhere; to say that spontaneous recovery accounts for it shifts the burden of explanation once again. Granted that spontaneous recovery is a time-dependent process (a position which is not without its critics, e.g., Skinner, 1950), we still must ask for some clue to the mechanism.[6]

In the light of these arguments, the spontaneous recovery hypothesis of PI

[6] The wheel has come full cycle now that some students of animal memory, pointing to the formal similarity between an experiment on PI and one on spontaneous recovery (e.g., first "go," then "no-go," then an interval and finally a test) are starting to think of the latter as a special case of PI (Spear, Chapter 2). Spontaneous recovery is here interpreted as forgetting which came last, reinforcement or nonreinforcement. This approach is very interesting and may indeed be correct, but of course once one has taken it to explain spontaneous recovery, one can no longer use spontaneous recovery to explain either PI or forgetting.

seems difficult to maintain. What then might account for the fact that PI increases with increasing retention interval? We suggest a simple possibility: forgetting. S forgets which of the two discriminations he has encountered more recently, an effect which grows as the retention interval increases and the relative recency of the two situations becomes more alike. This view fits in well with the finding that for verbal materials PI decreases with increasing interval between first-learned and second-learned tasks (Alin, 1968). In short, it may be that the explanation of PI rests with the phenomenon of forgetting and not the other way around.

Our position here has much in common with the analysis of proactive interference effects presented by Spear in this volume (Chapter 2). Spear starts with the assumption that the animal has a tendency to respond as it had done most recently. If the most recent of two conflicting reaction patterns was required only a short while ago, S has no problems; if the time interval is longer, S needs some additional retrieval cue to avoid confusion between the two opposed tasks. As evidence, Spear presents several studies in which rats learn a passive-avoidance task followed by an active one with tests after several retention intervals; the most impressive is the demonstration that a tone present during the first-learned task will augment PI if also present during the test. (In the light of our previous comments on the forgetting of incidental stimulus features, it is interesting to note that this effect reaches a maximum with a 4-hour retention interval. As Fig. 5 of Spear's chapter shows, while PI is still at a high level after 1 or 21 days, it is no longer enhanced by the specific retrieval cue.) On the other hand, Spear's position is not without difficulties. To him, proactive interference in animals operates essentially upon retrieval, hence the effect should be primarily on the first test trial and should dissipate very rapidly thereafter (in his terms, a "lapse" rather than a "loss"). This is not always the case. True enough, PI of runway performance produced by prior unreinforced trials is very evanescent (Spear, Hill, & O'Sullivan, 1965) as are the retention decrements of active avoidance produced by previously learned passive avoidance (Spear, Chapter 2). Our own work gave quite different results: In each of three different experiments, while PI declined noticeably it was still very much in evidence after some 30 test trials (Gleitman & Jung, 1963; Maier, Allaway, & Gleitman, 1967; Maier & Gleitman, 1967). Similar results were obtained by Koppenaal and Jagoda (1968) who found considerable PI in the relearning of a T-maze. It would seem that sometimes PI is more than a lapse. The same holds for (ordinary) forgetting. In virtually all of our studies in which we have found a retention decrement, the effect was obtained both for "recall" and "relearning" (e.g., Gleitman & Bernheim, 1963; Gleitman & Steinman, 1963, 1964).

   c. *Some Conflicting Evidence.* Whatever our intepretations of PI effects in animals, we must face the fact that such effects have not been obtained by all investigators who tested for PI in choice situations. While clear evidence for PI was found in six studies (Chiszar & Spear, 1968a; Gleitman & Jung, 1963; Koppenaal & Jagoda, 1968; Maier, Allaway, & Gleitman, 1967; Maier & Gleitman, 1967; Spitzner & Spear, 1967), two others showed no signs of it (Crowder, 1967; Rickard, 1965). In two further experiments (Kehoe, 1963; Spitzner & Spear, 1967), PI was not found when assessed by absolute number of errors but was obtained in the rather diluted form of "intrusion errors": Given that an error was made, it was more likely to be the response acquired in the first-learned task.

Many suggestions have been offered to explain failures to obtain PI in various choice situations: the use of noncorrection procedures, the employment of short retention intervals, overly high terminal acquisition levels of the second-learned task, negative transfer from the first-learned task, and distributed practice on either the first or second task. Unfortunately none of these can by themselves account for all of the findings. The use of the noncorrection method is surely not critical since several studies that have employed it yielded unambiguous PI (e.g., Maier & Gleitman, 1967). The use of short retention intervals or high terminal acquisition levels can also be ruled out: PI has been found after retention intervals as short as three days (e.g., Koppenaal & Jagoda, 1968) and following terminal acquisition levels of the second-learned task as high as 98.5% correct responding (Chiszar & Spear, 1968a). Nor is the absence of negative transfer from the first to the second task a guarantee of PI, for several studies testify otherwise (Crowder, 1967; Rickard, 1965). Finally, we can see no evidence to attribute the absence of PI in animal choice situations to distributed practice on either the first or the second task despite the fact that learning either task over several sessions is known to reduce PI for verbal materials (Keppel, 1964; Underwood & Ekstrand, 1966). Several investigators used distributed trials for both tasks; some found PI (e.g., Maier & Gleitman, 1967) while others did not (e.g., Kehoe, 1963). Others massed the second task; again, some found PI (e.g., Koppenaal & Jagoda, 1968) and some did not (e.g., Crowder, 1967).

Perhaps this account is not as damaging as it first appears since most of the studies cited differed in so many respects that comparison among them is rather difficult. Even so, one feels safe enough in concluding that none of the factors so far proposed can by itself account for the overall pattern of results. Future research should certainly be conducted more systematically, but more important, future hypotheses will have to be less vague and probably less simple than the ones offered to date.

B. CHANGE OF STIMULUS OR CONTEXT

A very different interpretation of retention decrements is that they are due to a change in the stimulus situation. If S learns some pattern of responses under one stimulus condition, his performance presumably will deteriorate if the stimuli (be they internal or external) are altered during the test. We have reason to suspect that something of this sort goes on from the everyday happenings of our own lives. We "forget" street names of cities we once lived in until we return for a later visit when it "all comes back in a flash"; we cannot think of a melody until someone whistles the first two notes; we have trouble recalling the name of an acquaintance if we meet him in surroundings different from those in which we usually see him. This makes good sense if we assume that at least some learning is associative. Given that A is associated to B, it is hardly surprising that B will be evoked more readily when A is provided than when it is absent. If one accepts this general approach one would expect that performance can be revived by "reinstating" part of the original learning situation. Such phenomena as warm-up (e.g., Irion, 1949) and savings during relearning, while not altogether unequivocal, still testify to the general plausibility of this position. More relevant to our present purposes is Spear's experimental demonstration of the importance of retrieval cues in animal retention (Spear, Chapter 2).

But even granting all this, why should forgetting increase with increasing retention interval? To account for this (so very basic) fact in terms of stimulus change, we must assume that the probability of stimulus change increases with time, an assumption rather similar to that made in Estes' (1955) treatment of spontaneous recovery and regression. This assumption suggests that forgetting should be less severe if original training occurred under varying rather than constant stimulus conditions: Having sampled a wider range of stimulus situations, the "broadly" trained Ss should be less likely to encounter a test condition that represents a serious change from training. We have tried to test this prediction in two experiments.

1. Varied vs. Constant Training in the Runway

The Ss were 36 female rats on a 22-hour food deprivation schedule and at 85% body weight. The apparatus was a standard runway, either black or white. Rats could be placed in the runway directly from the home cage (no restraint), or first placed in a restraint box for 30 seconds before the start of a trial (30-second restraint). If run on no restraint, rats were handled with the bare hand; if on 30-second restraint, they were handled with a heavy leather glove. At the beginning of the experiment, the animals were randomly assigned to either the black or the white runway, the no restraint or the 30-second restraint condition, and remained in these conditions for pretraining and four acquisition trials. The Ss were run daily, receiving one trial per day.

After the fourth trial, the animals were assigned to two equal groups, matched by their performance up to this point. Both groups were run for another 12 trials, again at one trial per day. Group C ("Constant") was run on constant stimulus conditions; both the runway color and the handling conditions were identical to those of the first four trials and remained thus for the remaining 12 trials. Group V ("Variable"), on the other hand, was run under varied stimulus conditions. For these animals both runway color and handling conditions were varied from trial to trial, starting with Trial 5. For example, if a particular animal in Group V had been run on the black runway and under no restraint for the first four trials, it would encounter a white runway and 30-second restraint on Trial 5. From that trial on, the conditions were varied from trial to trial alternating between the new pattern and the old (that is, the one that held for Trials 1–4) in accordance with a Gellerman order. This sequence continued from Trial 5 through Trial 16.

Both groups were tested after a retention interval of 56 days following Trial 16. For Group C the stimulus conditions on the test were identical to those encountered during acquisition. Half of the Ss in Group V were tested on the same conditions to which they had been assigned during Trials 1–4; the other half were tested on the opposite stimulus condition. (Of course, all Ss in Group V had encountered both stimulus conditions during Trials 5–16.) There were five test trials, administered once a day, and with identical stimulus conditions for all five. In all other respects, the procedure during the retention test was identical to that during acquisition.

The various subconditions within the two major groups (e.g., black vs. white runway, no restraint vs. 30-second restraint) had no systematic effects. The data for the subgroups were therefore combined and results are presented for the two main groups as such. Figure 11 presents mean total times (in log seconds) for the 16 trials of acquisition and the five trials of the retention test for Groups C and V. The figure indicates that the stimulus dimensions we varied were attended to: Group V shows a performance decrement on Trial 5 when the change was first introduced. It is also clear that there was forgetting: The performance scores on the first retention test are worse for both groups than those on the last day of acquisition. This effect holds for virtually all of the subjects. Comparing terminal acquisition performance (mean total time for Trials 14, 15, and 16) with performance on the first retention test, we found increased time scores on the retention test for 17 out of 18 Group C subjects and for 16 out of 18 Group V subjects. Obviously both groups forgot. The question is whether the subjects in Group V forgot less than those in Group C. Figure 11 suggests that they did, but the effect was not statistically significant ($t = 1.49$, $df = 34$, $p > .10$). Considering the rather large number of subjects run (18 per group), this negative result is of some importance; either there is no effect or, if it does exist, it is so minor as to be submerged in the statistical noise.

2. Varied vs. Constant Training in the Lever Box

A similar experiment was performed comparing the effect of constant and varied experience in the lever box. The Ss were 39 male rats, under the usual

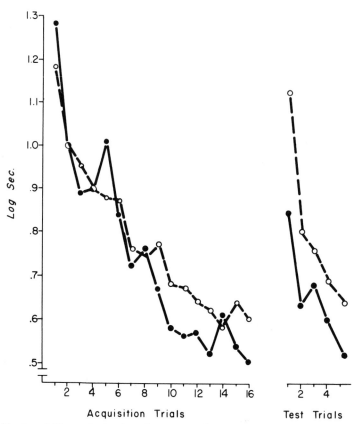

Fig. 11. Acquisition and retention of runway performance for groups run under constant (o - - o) and variable (●——●) stimulus conditions during acquisition (mean total times).

hunger drive schedule. At the beginning of the experiment all Ss were randomly assigned to one of four test chambers. Two of these chambers were equipped with a standard pellet feeder dispensing the usual Noyes pellets. In one of these, the lever was mounted on the right; in the other it was placed on the left. The other two test chambers were provided with dippers which could present a small amount (0.04 ml) of condensed milk; here too, the lever was mounted on the right for one of these chambers, on the left for the other. The levers were retractable, presented at the beginning of each trial and withdrawn as soon as the trial was ended.

The Ss were run with 30 discrete trials per session, and were required to press the lever five times on every trial to gain reinforcement and end the trial. For the first two sessions, all Ss were run in the chamber to which they were initially assigned. After the second session, they were assigned to two groups, matched by performance on Session 2 (which was indexed by median time per trial). Group C continued acquisition training under constant stimulus conditions, and remained in the same box for all future sessions. Subjects in Group V were run under varied conditions starting with Session 3; from session to session they were switched from one box to another (within sessions they stayed in the same box). The alternation pattern over sessions was determined by a Gellerman order. Both groups continued acquisition training for another five sessions. Following Session 7, the Ss within each group were assigned to two subgroups, matched by terminal acquisition performance. Two of the subgroups (C-1 and V-1) were tested 1 day after Session 7; the other two subgroups (C-60 and V-60) were tested 60 days later. The retention test followed the identical procedure in force during acquisition, except that half of the Ss in the V-groups were tested in the chamber they had worked in on the first two sessions, while the other half were tested in the chamber they were switched to on Session 3.

The results were analyzed using median time per trial for each S in blocks of 10 trials. Results during acquisition showed no systematic differences between the groups. The critical comparisons involve the first block on Session 8 (administered one day after Session 7 for Groups C-1 and V-1, and 60 days later for Groups C-60 and V-60). The means of the median time per trial on the first block of Session 8 were 8.7 seconds and 8.1 seconds for Groups C-1 and V-1, respectively; for Groups C-60 and V-60 the corresponding figures were 12.5 seconds and 10.1 seconds. Analysis of variance of these results yielded a highly significant main effect of retention interval ($F = 9.17$, $df = 1/35$, $p < .005$); evidently there is forgetting of the lever-pressing response. Neither the main effect of constant vs. variable training ($F = 2.58$, $df = 1/35$, $p > .10$) nor its interaction with retention interval ($F = 1.40$, $df = 1/35$, $p > .20$) approached statistical significance.

The outcome of this experiment is quite similar to that obtained with the runway. In both cases, a 2-month retention interval produced forgetting. In both cases there was a slight indication that the degree of forgetting was less when the training conditions were variable rather than constant, but in neither study was the effect large enough to get within reach of statistical acceptability. Such differences as were obtained are probably due to sampling error, but even if the effects turn out to be genuine after all, they are absurdly small. Forgetting is a major and well-nigh ubiquitous phenomenon. It is hard to believe that we have caught the essence of such a massive phenomenon with such a small and dubious effect.

Does this mean that stimulus change (be the stimulus external or internal, in the sense of set or context) has little or no relevance to forgetting? Hardly. We do not deny that performance decrement is often due to a change in the situation evoking the response.[7] What is at issue is whether this effect can account for retention loss which increases with increasing retention interval, specifically the assumption that stimulus change increases over time.

It is of course possible that our manipulations of stimulus conditions did not get at the critical stimulus dimensions, specifically those which by hypothesis change over time. This argument is hard to rebut though we can again point to the fact that in the runway study the animals showed serious performance decrement when the change was first introduced during training (see Fig. 11). This certainly indicates that the dimensions we varied were attended to; of course, it does not demonstrate that they were the ones that changed over the time interval.

C. DECAY

A third alternative is the decay theory of forgetting. This position asserts that the memory trace (or perhaps the access to this trace) becomes degraded by an unknown process that operates over time. This position has been largely ignored for almost three decades since McGeoch's critique of the law of disuse (McGeoch, 1932). It is often characterized as essentially vacuous (e.g., Hilgard & Bower, 1966, p. 496), and with some reason: In the absence of some further hypothesis that specifies the physiological mechanisms responsible for decay, one might well argue that decay theory amounts to little more than a restatement of the facts of the forgetting curve. But this argument is hardly compelling, especially since the evidence for the competitors of decay theory is so very meager. These competitors (especially the interference theories) are in one sense rather attractive. They offer an explanation for forgetting that requires no further physiological machinery beyond that already required for learning itself; perhaps this is one reason for their continued popularity. But the actual state of affairs may be quite different: The assumption that memory traces are permanent is no more plausible than the assertion that they (or perhaps their

---

[7] Recent investigators of animal memory have discovered an important fact that might perhaps be explained as a stimulus change effect of this kind: Infant rats forget much more quickly than adult ones, a phenomenon found both in aversive (Campbell & Campbell, 1962; Smith, 1968) and in appetitive learning situations (Campbell, Jaynes, & Misanin, 1968). Since the infant is growing rapidly one might suppose that the fairly long retention intervals typically used in these studies (of the order of a few weeks) would have drastically altered the stimulus situation as he perceives it; from his point of view, the runway must have shrunk.

There are many alternative interpretations some of which may well be more plausible (e.g., changes in the central nervous system). But we should notice that one factor is very unlikely to be the cause of the phenomenon, and that is PI: An infant rat should have acquired fewer extra-experimental responses than his older brothers because he had less time to acquire them.

access routes) gradually disintegrate. Under the circumstances we might do worse than to entertain decay theory seriously as a genuine alternative, even though the formulation of any specific hypotheses about the physiological mechanisms that underlie the decay process may not be possible at this time. Indeed, some investigators concerned with the biological basis of memory seem to have moved toward a decay hypothesis (e.g., Deutsch & Leibowitz, 1966). Furthermore, there has been a serious revival of interest in decay theory among students of short-term memory where the approach has met with some success (e.g., Broadbent, 1958; Brown, 1958).

It is important to recognize that the phenomena of experimentally induced interference are not inconsistent with a decay approach. Partial decay of a trace would not necessarily lead to failure of recall, assuming there was enough redundancy in the stored material. But if two similar traces are aroused (which presumably is the case if interfering material has been acquired) then partial information may no longer be adequate. This point has been elaborated by Brown (1958, p. 13) who points out that ". . . failure of discrimination, i.e. confusion between responses, cannot be regarded as a primary cause of forgetting. Failure of discrimination produces forgetting of that which determines which of the responses is correct. It is thus a possible *effect* of forgetting, however caused, but is not itself a primary cause of forgetting . . ."

There may be some first steps that can be taken to give decay theory some genuine, testable substance, even in the absence of any specific hypothesis about its physiological underpinnings. If the decay process is viewed as part and parcel of the normal biological life functioning of the organism (e.g., somehow related to metabolic mechanisms), then whatever speeds up these overall normal patterns (of which the hypothesized decay process is somehow a part) must necessarily increase forgetting; whatever slows them down, will slow down memory loss. This line of attack was first utilized by French (1942) who accelerated the forgetting of a maze habit in goldfish by heating them during the retention interval and who attributed his results to an increase in the rate of decay. Unfortunately, these results permit alternative interpretations. French employed a multiple maze with a somewhat obscure reinforcement condition (escape from confinement and the company of another goldfish in the goal chamber), utilized a rather small retention interval (22 hours), and had no control for warm-up. Furthermore, his study did not exclude the possibility that an ongoing consolidation process was disrupted by increasing the temperature. French's experimental animals were brought up to the high temperature within perhaps 30 minutes to an hour after learning. This period may still be within the range of the consolidation process in fish. Studies of consolidation in these animals indicate that, under some circumstances at least, the process can continue for much longer than had once been believed, certainly of the order of hours (Agranoff, Chapter 6).

P. Rozin and H. Gleitman (in an unpublished study) tried to control for some of these procedural problems, and ultimately to extend the phenomenon more widely. Unfortunately, we were unable to obtain any clear-cut evidence. Our studies used goldfish in an avoidance shuttle tank in an experimental paradigm patterned after French's. The fish were trained to a moderate criterion in a shuttle tank and then extinguished 1 day, 4 weeks, or 8 weeks after acquisition. The 4-week and 8-week retention intervals were spent in a tank at either 25–26°C (the training and test temperature) or 33°C. To avoid any possible disruption of consolidation processes, the "hot groups" remained at the lower temperatures for 2 days before transfer to the warmer temperature; to avoid complications with readaptation they were brought back to the training temperature 2 days before test. To control for the effects of prior temperature on learning and performance, two groups of naive fish were maintained for 8 weeks at 25–26°C and 33°C, respectively, and then trained on the avoidance task.

The results of a first study were in line with French's findings. Retention deteriorated with increasing interval, and with increasing temperature during that interval. The results of a second study which used several temperature levels during the interval were not in agreement. While we still found a large effect of retention interval, the temperature effect on retention was virtually absent (at best, limited to the first test-trial only and quickly declining as trials progressed). To date, we have not been able to resolve the inconsistency in the results of these two experiments.

Some other investigators pursuing a similar line of attack have been more successful. A recent study by Alloway (1969) found a facilitating effect of cold during the interval on the retention of a T-maze habit in the grain beetle (*Tenebrio molitor*). Related effects were found in two studies on goldfish who retained a visual discrimination better if treated with chlorpromazine (Rensch & Dücker, 1966) or kept in darkness (Dücker & Rensch, 1968) during the retention interval. Rensch and Dücker interpret both the chlorpromazine and the darkness effect as support for the decay position. They argue that the critical factor in trace breakdown is overall neural activity; whatever lowers the general metabolic level, or perhaps the overall excitatory state of the relevant system (e.g., lowered visual input during darkness, given that the relevant trace is visual) should slow up the forgetting process.

We should emphasize that interference theory can account for the slowing down of the forgetting rate by chlorpromazine, darkness, or for that matter cold temperatures, on the hypothesis that all of these minimize learning (and thus later interference) during the retention interval. This hypothesis is difficult to establish, and equally difficult to disconfirm. To establish it one must show, not only that the factor which accelerates forgetting (e.g., heat, light) speeds up learning during the retention interval, but also that the learned patterns which are acquired during this interval interfere with the habit the retention of which is being tested. To disconfirm the interference interpretation one must show that heat, light, and so on have no effect upon rate of learning during the retention interval. One approach might be to make learning during the retention interval virtually impossible, for example by anesthesia; should the temperature effect, for example, still hold under these conditions, an interference hypothesis becomes rather implausible. Such a study is far from easy to execute. (We have tried unsuccessfully to develop an appropriate technique for over a year.)

III. Summary

Our work shows quite clearly that animals are not immune to long-term forgetting. We know that this effect is sometimes caused by a failure of retrieval, but whether it is always best understood in this manner or whether it sometimes represents a genuine, irremediable loss is as yet a completely open question. To date, we know very little about the causes of forgetting, be it in men or in animals. But we hope at least to have demonstrated that some generally shared beliefs about forgetting are rooted more deeply in conviction than in factual evidence. In particular, we have found no evidence to support either an interference or a stimulus-change theory of long-term forgetting. While we have not been able to provide any positive evidence for decay theory, we nevertheless feel that this long-neglected theoretical alternative deserves more serious exploration.

ACKNOWLEDGMENTS

The author is deeply indebted to several collaborators who have participated on various phases of this project, particularly to Fredda Steinman, Steven F. Maier, and Paul Rozin. Further thanks are due to Werner K. Honig and Henry James for a critical discussion of the manuscript that was far beyond the normal call of editorial duty.

REFERENCES

Aiken, E. G. & Gibson, K. L. Continuous and fixed ratio reinforcement effects in extinction one day and three weeks after acquisition. *Psychonomic Science, 1965,* 3, 527–528.

Alexander, D. S. Retention of the partial reinforcement effect in rats. Unpublished first-year research report, University of Pennsylvania, 1966.

Alin, L. H. Proactive inhibition as a function of the time interval between the learning of the two tasks and the number of prior lists. *Journal of Verbal Learning and Verbal Behavior,* 1968, 7, 1024–1029.

Alloway, T. M. Effects of low temperature upon acquisition and retention in the grain beetle (*Tenebrio molitor*). *Journal of Comparative and Physiological Psychology,* 1969, 69, 1–8.

Brady, J. V. The effect of electro-convulsive shock on a conditioned emotional response: The permanence of the effect. *Journal of Comparative and Physiological Psychology,* 1951, 44, 507-511.

Briggs, G. E. Acquisition, extinction, and recovery functions in retroactive inhibition. *Journal of Experimental Psychology,* 1954, 47, 285-293.

Broadbent, D. E. *Perception and communication.* Oxford: Pergamon Press, 1958.

Brown, J. Some tests of the decay theory of immediate memory. *Quarterly Journal of Experimental Psychology,* 1958, 10, 12-21.

Brush, F. R. Retention of aversively motivated behavior. In F. R. Brush (Ed.), *Aversive conditioning and learning.* New York: Academic Press, 1971, in press.

Brush, F. R., Myer, J. S., & Palmer, M. E. Effects of kind of prior training and intersession interval upon subsequent avoidance learning. *Journal of Comparative and Physiological Psychology,* 1963, 56, 539-545.

Bunch, M. E. A comparison of retention and transfer of training from similar material after relatively long intervals of time. *Journal of Comparative Psychology,* 1941, 32, 217-231.

Campbell, B. A., & Campbell, E. H. Retention and extinction of learned fear in infant and adult rats. *Journal of Comparative and Physiological Psychology,* 1962, 55, 1-8.

Campbell, B. A., Jaynes, J., & Misanin, J. R. Retention of a light-dark discrimination in rats of different ages. *Journal of Comparative and Physiological Psychology,* 1968, 66, 467-472.

Chiszar, D. A., & Spear, N. E. Proactive interference in a T-maze brightness-discrimination task. *Psychonomic Science,* 1968, 11, 107-108. (a)

Chiszar, D. A., & Spear, N. E. Proactive interference in retention of nondiscriminative learning. *Psychonomic Science,* 1968, 12, 87-88. (b)

Crespi, L. P. Quantitative variation of incentive and performance in the white rat. *American Journal of Psychology,* 1942, 55, 467-517.

Crowder, R. G. Proactive and retroactive inhibition in the retention of a T-maze habit in rats. *Journal of Experimental Psychology,* 1967, 74, 167-171.

Denny, M. R., & Ditchman, M. E. The locus of maximal "Kamin Effect" in rats. *Journal of Comparative and Physiological Psychology,* 1962, 55, 1069-1070.

Desiderato, O., Butler, B., & Meyer, C. Changes in fear generalization gradients as a function of delayed testing. *Journal of Experimental Psychology,* 1966, 72, 678-682.

Deutsch, J. A., & Leibowitz, S. F. Amnesia or reversal of forgetting by anticholinesterase, depending simply on time of injection. *Science,* 1966, 153, 1017-1018.

Dücker, G., & Rensch, B. Verzögerung des Vergessens erlernter visuellen Aufgaben bei Fischen durch Dunkelhaltung. *Pfluegers Archiv fuer die Gesamte Physiologie des Menschen und der Tiere,* 1968, 301, 1-6.

Ebbinghaus, H. *Das Gedächtnis: Untersuchungen zur experimentellen Psychologie.* Leipzig: Duncker & Humbolt, 1885.

Ehrenfreund, D., & Allen, J. Perfect retention of an instrumental response. *Psychonomic Science,* 1964, 1, 347-348.

Ekstrand, B. R., & Underwood, B. J. Free learning and recall as a function of unit-sequence and letter-sequence interference. *Journal of Verbal Learning and Verbal Behavior,* 1965, 4, 390-396.

Ellson, D. G. Quantitative studies of the interaction of simple habits. I. Recovery from specific and generalized effects of extinction. *Journal of Experimental Psychology,* 1938, 23, 339-358.

Estes, W. K. Statistical theory of spontaneous recovery and regression. *Psychological Review,* 1955, 62, 145-154.

Finger, F. W. Retention and subsequent extinction of a simple running response following varying conditions of reinforcement. *Journal of Experimental Psychology*, 1942, **31**, 120–133.

French, J. W. The effect of temperature on the retention of a maze habit in fish. *Journal of Experimental Psychology*, 1942, **31**, 79–87.

Gagné, R. M. The retention of a conditioned operant response. *Journal of Experimental Psychology*, 1941, **29**, 296–305.

Gleitman, H., & Bernheim, J. W. Retention of fixed-interval performance in rats. *Journal of Comparative and Physiological Psychology*, 1963, **56**, 839–841.

Gleitman, H., & Holmes, P. Retention of incompletely learned CER in rats. *Psychonomic Science*, 1967, **7**, 19–20.

Gleitman, H., & Jung, L. Retention in rats: The effect of proactive interference. *Science*, 1963, **142**, 1683–1684.

Gleitman, H., & Steinman, F. Retention of runway performance as a function of proactive interference. *Journal of Comparative and Physiological Psychology*, 1963, **56**, 834–838.

Gleitman, H., & Steinman, F. Depression effect as a function of retention interval before and after shift in reward magnitude. *Journal of Comparative and Physiological Psychology*, 1964, **57**, 158–160.

Gleitman, H., Steinman, F., & Bernheim, J. W. Effect of prior interference upon retention of fixed-interval performances in rats. *Journal of Comparative and Physiological Psychology*, 1965, **59**, 461–462.

Gonzalez, R. C. Influence of contrasted reward and nonreward on instrumental behavior. Paper presented at symposium on the role of reward and nonreward in instrumental learning, Haverford College, 1970.

Gonzalez, R. C., & Bitterman, M. E. Spaced-trials partial reinforcement effect as a function of contrast. *Journal of Comparative and Physiological Psychology*, 1969, **67**, 94–103.

Gonzalez, R. C., Gleitman, H., & Bitterman, M. E. Some observations on the depression effect. *Journal of Comparative and Physiological Psychology*, 1962, **55**, 578–581.

Gossette, R. L., & Hood, P. The reversal index (RI) as a joint function of drive and incentive level. *Psychonomic Science*, 1967, **8**, 217–218.

Grissom, R. J., Suedfeld, P., & Vernon, J. Memory for verbal material: Effects of sensory deprivation. *Science*, 1962, **138**, 429–430.

Hilgard, E. R., & Atkinson, R. C. *Introduction to psychology.* New York: Harcourt, Brace, 1967.

Hilgard, E. R., & Bower, G. H. *Theories of learning.* New York: Appleton, 1966.

Hill, W. F., Cotton, J. W., Spear, N. E., & Duncan, C. P. Retention of T-maze learning after varying intervals following partial and continuous reinforcement. *Journal of Experimental Psychology*, 1969, **79**, 584–585.

Hoffman, H. S., Fleshler, M., & Jensen, P. Stimulus aspects of aversive controls: The retention of conditioned suppression. *Journal of the Experimental Analysis of Behavior*, 1963, **6**, 575–583.

Honig, W. K. Attentional factors governing the slope of the generalization gradient. In R. M. Gilbert & N. S. Sutherland (Eds.), *Animal discrimination learning.* New York: Academic Press, 1969. Pp. 35–62.

Houston, J. P. Proactive inhibition and competition at recall. *Journal of Experimental Psychology*, 1967, **75**, 118–121.

Irion, A. L. Retention and warming-up effects in paired-associate learning. *Journal of Experimental Psychology*, 1949, **39**, 669–675.

Jenkins, J. G., & Dallenbach, K. M. Obliviscence during sleep and waking. *American Journal of Psychology*, 1924, **35**, 605–612.

Jones, M. B. Effect, change, and expectation of reward. *Psychological Review,* 1952, **59,** 227–233.

Kamin, L. J. The retention of an incompletely learned avoidance response. *Journal of Comparative and Physiological Psychology,* 1957, **50,** 457–460.

Kamin, L. J. Temporal and intensity characteristics of the conditioned stimulus. In W. F. Prokasy (Ed.), *Classical conditioning.* New York: Appleton, 1965. Pp. 118–147.

Kehoe, J. Effects of prior and interpolated learning on retention in pigeons. *Journal of Experimental Psychology,* 1963, **65,** 537–545.

Kendler, H. H., & Lachman, R. Habit reversal as a function of schedule of reinforcement and drive strength. *Journal of Experimental Psychology,* 1958, **55,** 584–591.

Keppel, G. Facilitation in short- and long-term retention of paired associates following distributed practice in learning. *Journal of Verbal Learning and Verbal Behavior,* 1964, **3,** 91–111.

Keppel, G. Retroactive and proactive inhibition. In T. R. Dixon and D. L. Horton (Eds.), *Verbal behavior and general behavior theory.* Englewood Cliffs, N.J.: Prentice-Hall, 1968. Pp. 172–213.

Kimble, G. A. *Hilgard and Marquis' conditioning and learning.* New York: Appleton, 1961.

Kirby, R. H. Acquisition, extinction, and retention of an avoidance response in rats as a function of age. *Journal of Comparative and Physiological Psychology,* 1963, **56,** 158–162.

Klein, S. B., & Spear, N. E. Forgetting by the rat after intermediate intervals ("Kamin Effect") as retrieval failure. *Journal of Comparative and Physiological Psychology,* 1970, **71,** 165–170.

Koppenaal, R. J. Time changes in the strengths of A-B, A-C lists; spontaneous recovery? *Journal of Verbal Learning and Verbal Behavior,* 1963, **2,** 310–319.

Koppenaal, R. J., & Jagoda, E. Proactive inhibition of a maze position habit. *Journal of Experimental Psychology,* 1968, **76,** 664–668.

McAllister, D. E., & McAllister, W. R. Forgetting of acquired fear. *Journal of Comparative and Physiological Psychology,* 1968, **65,** 352–355.

McAllister, W. R., & McAllister, D. E. Increase over time in the stimulus generalization of acquired fear. *Journal of Experimental Psychology,* 1963, **65,** 576–582.

McGeoch, J. A. Forgetting and the law of disuse. *Psychological Review,* 1932, **39,** 352–370.

McGeoch, J. A. *The psychology of human learning.* New York: Longmans, Green, 1942.

Magdsick, W. K. The curve of retention of an incompletely learned problem in albino rats at various age levels. *Journal of Psychology,* 1936, **2,** 25–48.

Maier, S. F., Allaway, T. A., & Gleitman, H. Proactive inhibition in rats after prior partial reversal: A critique of the spontaneous recovery hypothesis. *Psychonomic Science,* 1967, **9,** 63–64.

Maier, S. F., & Gleitman, H. Proactive interference in rats. *Psychonomic Science,* 1967, **7,** 25–26.

Melton, A. W., & Irwin, J. M. The influence of degree of interpolated learning on retroactive inhibition and the overt transfer of specific responses. *American Journal of Psychology,* 1940, **53,** 173–203.

Minami, H., & Dallenbach, K. M. The effect of activity upon learning and retention in the cockroach, *(Periplaneta americana). American Journal of Psychology,* 1946, **59,** 1–58.

Mote, F. A., Jr., & Finger, F. W. The retention of a simple running response after varying amounts of reinforcement. *Journal of Experimental Psychology,* 1943, **33,** 317–322.

Perkins, C. C., Jr., & Weyant, R. G. The interval between training and test trials as determiner of the slope of generalization gradients. *Journal of Comparative and Physiological Psychology,* 1958, **51,** 596–600.

Postman, L. Extra-experimental interference and the retention of words. *Journal of Experimental Psychology,* 1961, **61**, 91–110.

Postman, L., & Rau, L. Retention as a function of the method of measurement. *University of California Publications in Psychology,* 1957, **8**, 217–270.

Postman, L., & Stark, K. Role of response availability in transfer and interference. *Journal of Experimental Psychology,* 1969, **79**, 168–177.

Rensch, B., & Dücker, G. Verzögerung des Vergessens erlernter visuellen Aufgaben bei Tieren durch Chlorpromazin. *Pfluegers Archiv fuer die Gesamte Physiologie des Menschen und der Tiere,* 1966, **289**, 200–214.

Rickard, S. Proactive inhibition involving maze habits. *Psychonomic Science,* 1965, **3**, 401–402.

Russell, R. W., & Hunter, W. S. The effects of inactivity produced by sodium amytal on the retention of the maze habit in the albino rat. *Journal of Experimental Psychology,* 1937, **20**, 426–436.

Skinner, B. F. Are theories of learning necessary? *Psychological Review,* 1950, **57**, 193–216.

Slamecka, N. J. Supplementary report: A search for spontaneous recovery of verbal association. *Journal of Verbal Learning and Verbal Behavior,* 1966, **5**, 205–207.

Smith, N. Effects of interpolated learning on the retention of an escape response in rats as a function of age. *Journal of Comparative and Physiological Psychology,* 1968, **65**, 422–426.

Spear, N. E. Retention of reinforcer magnitude. *Psychological Review,* 1967, **74**, 216–234.

Spear, N. E., Hill, W. F., & O'Sullivan, D. J. Acquisition and extinction after initial trials without reward. *Journal of Experimental Psychology,* 1965, **69**, 25–29.

Spear, N. E., & Spitzner, J. H. Effect of initial nonrewarded trials: Factors responsible for increased resistance to extinction. *Journal of Experimental Psychology,* 1967, **74**, 525–537.

Spear, N. E., & Spitzner, J. H. Residual effects of reinforcer magnitude. *Journal of Experimental Psychology,* 1968, **77**, 135–149.

Spence, K. W. *Behavior theory and conditioning.* New Haven, Conn.: Yale Univ. Press, 1956.

Spence, K. W. A theory of emotionally based drive (D) and its relation to performance in simple learning situations. *American Psychologist,* 1958, **13**, 131–141.

Spence, K. W., Farber, I. E., & McFann, H. H. The relation of anxiety (drive) level to performance in competitional and noncompetitional paired-associates learning. *Journal of Experimental Psychology,* 1956, **52**, 296–305.

Spitzner, J. H., & Spear, N. E. Studies of proactive interference in retention of the rat. Paper presented at the meetings of the Psychonomic Society, St. Louis, 1967. Reported in N. E. Spear, Chapter 2, this volume.

Steinman, F. Retention of alley brightness in the rat. *Journal of Comparative and Physiological Psychology,* 1967, **64**, 105–109.

Thomas, D. R., & Burr, D. E. S. Stimulus generalization as a function of the delay between training and testing procedures: A reevaluation. *Journal of the Experimental Analysis of Behavior,* 1969, **12**, 105–109.

Thomas, D. R., & Lopez, L. J. The effects of delayed testing on generalization slope. *Journal of Comparative and Physiological Psychology,* 1962, **55**, 541–544.

Thomas, D. R., Ost, J., & Thomas, D. Stimulus generalization as a function of the time between training and testing procedures. *Journal of the Experimental Analysis of Behavior,* 1960, **3**, 9–14.

Tsai, C. A comparative study of retention curves for motor habits. *Comparative Psychology Monographs,* 1924, **2**, 11.

Underwood, B. J. "Spontaneous recovery" of verbal associations. *Journal of Experimental Psychology,* 1948, **38**, 429–439.

Underwood, B. J. Proactive inhibition as a function of time and degree of prior learning. *Journal of Experimental Psychology,* 1949, **39,** 24–34.

Underwood, B. J. Interference and forgetting. *Psychological Review,* 1957, **64,** 49–60.

Underwood, B. J., & Ekstrand, B. R. An analysis of some shortcomings in the interference theory of forgetting. *Psychological Review,* 1966, **73,** 540–549.

Underwood, B. J., & Postman, L. Extraexperimental sources of interference in forgetting. *Psychological Review,* 1960, **67,** 73–95.

Van Ormer, E. B. Retention after intervals of sleep and of waking. *Archives of Psychology, New York,* 1932, **21,** No. 137.

Wagner, A. R. Effects of amount and percentage of reinforcement and number of acquisition trials on conditioning and extinction. *Journal of Experimental Psychology,* 1961, **62,** 234–242.

Youtz, R. E. P. The change with time of a Thorndikian response in the rat. *Journal of Experimental Psychology,* 1938, **23,** 128–140.

Chapter 2

FORGETTING AS RETRIEVAL FAILURE[1]

Norman E. Spear

In this chapter the view is taken that forgetting has either of two sources: (1) a failure to maintain the storage of information which has been acquired, or (2) an inability to retrieve this information from storage. It has become customary to overemphasize the part which storage plays in animal memory and to neglect the importance, both empirical and theoretical, which the process of retrieval plays in determining retention by animals. This is due to a variety of factors: The difficulty of studying the hypothetical "consolidation" process in man and the comparative ease of doing so in animals; the pervasive influence of verbal instructions in human retention, the consequent emphasis on the selectiveness of forgetting in man and the difficulty of manipulating the analagous central processes in animals; and the apparent ease with which nearly total recall may often be demonstrated at the subhuman level. I believe that the understanding of memory might be more effectively served by an alternative view emphasizing the role of retrieval in animal memory. This view is expressed as the thesis of this chapter: *If equivalent memory storage is once attained, subsequent differences in retention are caused by retrieval differences.*

However, my purpose is not to present a case for retrieval failure as a source of forgetting to the unqualified exclusion of decay of storage. I have the more modest intention of attributing some diverse cases of retention loss to a common source, namely retrieval failure. Surely the existence of memory decay in some form is quite plausible, although perhaps limited to chemical or anatomical changes in brain sites linked with memory, and insignificant in terms of

[1] Preparation of this report was supported by a Rutgers University Research Council Faculty Fellowship and by the following grants from the National Institute of Mental Health: MH 0888, MH 12064, MH 18619, and Research Scientist Development Award MH 47359. This report has profited from work conducted in our laboratory by David Chiszar, David Feigley, Stephen Klein, Ira Landau, Ralph Miller, Jeffrey Nagelbush, Patricia Parsons, Edward Riley, and Joseph Spitzner, the technical advice and assistance of Norman Richter, and the secretarial skills of Mrs. L. Bruska and Mrs. M. Reich.

behavior. In any event, should failure of storage be decisively shown to influence behavior, only the complexity of memory theories would be affected while the relevance of retrieval processes would be undiminished.

If we are to study retrieval processes, experimental control is required over the amount of information originally stored. Such control is no problem so long as experimental treatments are not introduced until after original storage has been assured. But the latter is not always possible, a circumstance most obvious in manipulation of subject variables such as age. Therefore, we regularly attempt to equate quantity of information originally stored—degree of original learning—by the combined use of multitrial training and independent groups of rats tested after at least two retention intervals, one group tested after a short interval and one after a long interval. The former permits approximation of equivalent learning among groups treated differently by bringing them to equivalent criteria of performance. The latter provides increased assurance of equivalent learning if equivalent performance also can be demonstrated on a retention test given after a short interval (cf. Underwood, 1964).

The inclusion of independent retention tests at more than one interval has another function which is particularly important for evaluating the role of the retention interval in retrieval failure. Surely the demonstration of previously learned performance at a retention test depends at least somewhat upon the degree to which circumstances of the retention test correspond to those of original learning. It is equally undeniable that E cannot reinstate the identical circumstances of learning at the retention test. Because of this inevitable stimulus change between training and testing, the potential for generalization decrement in performance is equally inevitable. Therefore, to estimate properly the influence of the retention interval independently of uncontrollable stimulus changes the difference in performance between independent groups of Ss given nominally identical short- and long-interval retention tests must be used.

The equation of certain sources of stimulus change in this way is immensely more important for animals than humans. If Ss are verbal, they may be reassured by implicit or explicit instructions that such uncontrolled stimulus change is inconsequential—"*today* you should try to remember the words you saw *yesterday*"—and they may be told precisely what class of events they are to remember. But the animal's sole source of assurance that the reinforcement contingencies have not changed since his most recent experimental experience is the degree of correspondence between contemporary stimuli and those of his previous experimental experience. Furthermore this consequence of language has been reflected by a concensus in methodology: Retention by humans is typically assessed by how well Ss remember what they are asked to remember, while retention by lower animals has most often been assessed by measuring how well the animal does what he last learned to do.

One final introductory note: Surely it is understood that analogies comparing retention of verbal learning and learning by nonverbal animals may be grossly misleading. The study of animal memory would flounder forever as an offshoot from theories of verbal memory. However, disciplines such as verbal learning have been aided in the past by the judicious use of constructs developed in the conditioning area. Similarly, information potentially useful to the study of animal memory may be gleaned from both methodological and theoretical ideas produced in the field of human learning, either "nonverbal" (e.g., Bilodeau, 1966) or verbal (e.g., Underwood, 1964, 1969). We must not allow the disjunction which exists between specialized subareas of learning in terms of jargon, phenomena, and theory to cause us to misjudge or ignore the merit of an idea because of its source.

I. Theoretical Framework

The purpose of this chapter is to suggest a means by which a collection of apparently diverse retention phenomena may be synthesized. The integrating device I have chosen emphasizes the importance of subtle stimulus changes between training and testing, thus centering on aspects of retrieval rather than initial learning or subsequent storage of memory. For the most part, the behavioral phenomena to be integrated involve failure of a rat to show reasonably well-learned behavior as a consequence of an increasing retention interval. A working assumption will be that the memory representing previously established learning does not fade or decay spontaneously over time; rather the memory remains potentially available, and if it is retrieved it may influence subsequent behavior.

The ideas which I have presented in this chapter have a great deal in common with those of Underwood (1969) who thinks of a memory as having many dimensions (see also Bower, 1967b). A memory is not simply a bonded stimulus–response but rather a collection of defining attributes (e.g., temporal, spatial, frequency, and modality). The concept of memory as multidimensional is applied as an explanatory device in this chapter as are some of the specific attributes suggested by Underwood.

Of course, certain of the attributes identified by Underwood seem to concern only verbal memory and have little relevance for understanding retention by animals. However, the basic principle—that an attribute represented by a specific retrieval cue may function to discriminate or retrieve a target attribute—is not so restricted. Moreover, this sort of reasoning is not unlike that underlying the stimulus-fluctuation model of forgetting which is neutral with regard to verbal capacity (e.g., Bower, 1967a; Estes, 1955, 1959).

Essentially, the present point of view may be summarized in the following way: (1) A response reflecting an acquired memory may be manifested if the relevant memory attribute is retrieved. (2) Retrieval of this target attribute—we may call it the "response attribute" for now—will occur upon the arousal of a sufficient number of associated memory attributes. (3) The primary elicitation of these associated attributes is caused by retrieval cues which are sufficiently similar to stimuli noticed by the animal during original acquisition of the memory. Thus the fundamental requirement for retrieval of a target memory attribute is that a sufficient number of these retrieval cues be present and noticed by the animal during the retention test.

We shall refer to two general classes of events which appear to function as retrieval cues, and hence memory attributes, for rats presented with instrumental learning tasks. One class may be termed "external stimulus events" and includes, for example, conditions and contingencies of reinforcement, discriminative and contextual stimuli, and specific sequences of antecedent stimuli. The second class may be termed "internal stimulus events" and includes, for example, the systemic chemical state of the organism and especially the chemical state of certain portions of the CNS, and the kinesthetic feedback from preceding responses. It should be clear that this classification scheme is arbitrary and a matter of convenience, having no significance as yet for evaluation of the present point of view.

Finally, it should be clear that the fundamental purpose of this chapter is not to identify specific stimulus events which function as effective memory attributes for animals. This is not to deny that some degree of such cataloging will be necessary to evaluate the effectiveness of any view of retention in terms of retrieval. Moreover, this matter has contemporary relevance for distinguishing among a variety of learning theories (e.g., Capaldi, 1966, 1967, and Chapter 3, compared with Amsel, 1962, 1967; Lewis, 1969, compared with McGaugh, Chapter 5). However, the primary aim of the research to be described here is to evaluate the role of changes in effective memory attributes and associated retrieval cues between original learning and the retention test, particularly in relation to the length of the retention interval.

II. Measures of Forgetting: "Lapses" and "Losses"

A short comment must be interjected concerning two distinct ways of measuring retention failure. A "lapse" is defined as an initial and transient failure to perform in accord with the most recent learning and is measured *before* reexposure to the most recent reinforcement conditions and contingencies. The lapse may be measured on the first trial following the retention interval or perhaps in terms of transfer to different reinforcement conditions such as extinction. The other measure, "loss," which is in one sense a misnomer since it is not meant to imply a permanent absence of memory, is defined as decrement in performance of the most recent learning *given* reinstatement of the most recent reinforcement conditions and contingencies (e.g., a decrement in the rate of relearning).

Certain data suggest that these measures may be dissociated empirically (Spear, 1969), so they may represent different kinds of forgetting, although we currently prefer to treat them as representing different levels of retrieval failure. Specific examples of this empirical dissociation include the following: (1) Greater long-term forgetting of one-way active avoidance by young than by older rats has been found reliably with the loss measure but not in terms of lapse (Feigley & Spear, 1970). (2) Effects of proactive interference (PI) on retention of go, no-go behavior, which readily occur in terms of lapse, have not been obtained in terms of the loss measure. (3) Certain aspects of the "Kamin Effect" obtained in terms of the loss measure do not appear in terms of lapse.

The bulk of the data to support these suggestions has come from our laboratory, although supporting data also may be found in existing literature (e.g., Carlton, 1969, pp. 303–304). Throughout this chapter a decrement in performance after a retention interval will be referred to as a retention (or memory) "lapse" or a retention (or memory) "loss," rather than simply "forgetting."

This concern for the measurement of retention is more than methodological busy work. In fact, the present point of view implies that the retention test not only assesses, but determines what is "remembered." We have asserted that retrieval of a target memory attribute depends upon the presence of certain cues at the retention test which activate other attributes of the memory. Some of the more important attributes for retrieval of a memory are those representing conditions of reinforcement (e.g., quality, quantity, and schedule of occurrence of the reinforcer) and aspects of the contingencies of reinforcements (e.g., where and under what circumstances the reinforcer occurs). Now suppose a treatment is introduced which increases "forgetting," presumably by eliminating certain retrieval cues or by reducing the effectiveness of others (e.g., by introducing cues similar to those represented by both the critical memory attributes and attributes of different memories). Although this treatment may cause a sizeable retention lapse, reinstatement of the conditions and the contingencies of reinforcement on subsequent relearning trials may activate a sufficient number of additional attributes to permit equivalent memory retrieval—i.e., equally good retention—for those Ss which have and those which have not been given the "forgetting" treatment. We soon shall have occasion for detailed descriptions of more specific interactions between treatments which influence retention and the nature of the retention test.

III. Modifications of Interference with Increasing Retention Intervals: Proactive Interference

The distinction between lapse and loss is especially pertinent for the analysis of PI effects. When a rat first learns to emit response A consistently and subsequently learns instead to emit response B consistently in the same situation, proactive interference from A is said to be in evidence when the rat later shows a smaller probability of giving response B than another rat which had learned only B.

Three points concerning our use of this definition require comment: (1) We have noted that the original degree of learning B must be equated across experimental and control conditions, and that both an immediate and a later test are required to evaluate the effect of the retention interval. Thus, for most experiments, these conditions require that the influence of PI on retention be exhibited as greater retention deficit attributable to the prior learning of A at the later, compared to the immediate, test.

(2) When the magnitude of the retention lapse for B increases over an interval more rapidly given A than not given A, we shall rather arbitrarily describe this as increasing PI. It may be objected that since S is not signalled—"instructed"—concerning the task that is required, S may "really" be behaving in accord with decreasing retroactive interference. However, I am concerned here fundamentally with the interaction of conflicting memories over retention intervals of relative inactivity, and accordingly, view this more or less semantic matter of RI vs. PI as unimportant at this stage.

(3) We shall see that PI rarely has been found very effective in terms of the "loss" response measure. Rather, its influence on retention occurs most profoundly before reinstatement of the reinforcement conditions and contingencies and disappears rapidly thereafter. This does not imply that the influence of PI on animal behavior is of little importance. On this basis it would also have to be concluded, somewhat paradoxically, that PI does not have an important effect on human forgetting, because the verbal learning literature is virtually unanimous in showing that PI affects forgetting very little after the recall trial. Typically, human Ss are shown the correct response subsequent to each recall anticipation, and these operations are in principle similar to reinstating the reinforcement conditions and contingencies for animal Ss.

A. Generality of the Influence of Proactive Interference on Animals

Retention lapse due to PI is a common phenomenon in animal behavior although it is not always treated as such. This may be demonstrated by the three paradigmatic cases shown in Table I. These represent simple and common effects which could be understood as PI effects if the control groups shown in Table I were included. While these cases have received some explicit study concerned with interference and forgetting, they are most widely observed without explicit control groups and under conditions in which the influence of the retention interval is only inferred. When the control groups are omitted some version of this behavior is familiar to most investigators of animal learning. Therefore, these

Table I

Experimental Paradigms Yielding More Rapid Forgetting Because of Prior Conflicting Learning–PI in Retention by Animals

Paradigm	Condition[a]	Prior learning	Most recent learning	Retention test
1.	PI	Traverse runway (rapidly) for food	Traverse runway (slowly) for no food (i.e., extinction)	Retention test trial (soon after most recent learning or later)
	C	None (but equivalent handling)		
2.	PI	Traverse runway (rapidly) for large palatable food reinforcer	Traverse runway (slowly) for unpalatable or small food reinforcer	Retention test trial (soon, or later)
	C	None (but equivalent handling)		
3.	PI	Traverse runway (rapidly) for food	Traverse runway (slowly), punished (shocked) in goal box	Retention test trial (soon, or later)
	C	None (but equivalent handling)		

[a]PI, proactive interference; C, control.

simple effects may be briefly described without detailed reference to our relevant unpublished data and with only a few examples of published references.

Perhaps the most familiar "retention" phenomenon with lower Ss that may be considered a lapse is spontaneous recovery. Suppose spontaneous recovery is treated as a failure to "remember" extinction because of PI from acquisition (cf. Bower, 1967a; Estes, 1955; Razran, 1939; Skinner, 1950). This requires a control group given only "extinction" as in the first experimental paradigm shown in Table I. These Ss also may run faster the longer the subsequent retention interval, but the increase is not as great as for Ss given prior rewarded trials. When this comparison is made, the effect of spontaneous recovery fits the pattern of a lapse; it is a relatively transient failure to do what S did just before the retention interval, and the magnitude of this failure increases with time. The "reverse" of spontaneous recovery also has been found. If nonrewarded trials are given first accompanied by slow running, followed by rewarded trials and fast running, rats run more slowly on the first retention trial the longer the retention interval (e.g., Spear, Hill, & O'Sullivan, 1965). Again the control group is provided by rats given only the second task (rewarded trials) prior to the retention test.

A similar case may be made for the effect of a previous reinforcer on retention of the influence of a different reinforcer (cf. Spear, 1967). Consider an experiment in which rats first ran to plain wet mash and then to wet mash adulterated with distasteful quinine, as in the second paradigm in Table I. The longer the interval after tasting the adulterated mixture, the less likely Ss were to run as if quinine were still present in the mash (i.e., slowly) compared to control groups given only the adulterated mixture (Chiszar & Spear, 1968a). Similarly, when the wet mash adulterated with quinine is given first followed by unadulterated mash, D. Chiszar and the author have found a corresponding lapse of the tendency to run fast on the first trial, given a sufficient retention interval. Moreover, to complement this effect following shifts in the quality of a reinforcer, a corresponding lapse in retention for the latest reinforcer has been found following shifts in the quantity of a reinforcer (e.g., Mellgren, 1970; Spear & Spitzner, 1968). In short, if animals are given successive learning experiences involving two different reinforcers, a lapse in the tendency to behave in accord with the most recent reinforcer is directly related to the length of the retention interval, and the lapse is greater than that found after experience with only one reinforcer.

A third case of retention lapse caused by prior learning may be found in terms of recovery from punishment (third paradigm, Table I). This recovery occurs over a retention interval of relative inactivity and should not be confused with recovery from punishment in the course of continued punishment. First, S is trained to go down an alley for food in a goal box. On some subsequent trial, S is given a shock in the goal box. If tested very soon afterward, S is reluctant to move toward the food. But if sufficient time passes, say an hour, S becomes less reluctant to do so (see also Bintz, Braud, & Brown, 1970). This effect is greater for rats trained first to run for food without shock than for rats that receive only shock trials without initial food-only trials.

These examples of the influence of prior learning on retention performance may be expanded to a variety of circumstances in which animals learn two or more conflicting tasks in succession prior to a test. In doing so they help implicate the basic effect as a substantial portion of the retention changes that take place in lower animals over intervals of relative inactivity. Nevertheless it would be misleading to maintain that these are uniformly robust phenomena. In fact, these phenomena seem surprisingly sensitive to subtle variations in controlling parameters which, in turn, are not well understood (e.g., Spear & Spitzner, 1967, 1968). Largely for this reason we have searched for a different experimental context in which to examine the effect of proactive interference on retention of go, no-go behavior.

B. PRELIMINARY ANALYSIS OF THE RETENTION LAPSE

1. General Procedure

The task that we have found useful for this work is simple one-way avoidance. Aversive conditioning has several advantages for studying retention. The most obvious is that it avoids confounding the effect of retention interval with variations in deprivation–depletion levels. Although necessary, it is most difficult to control the latter when retention is studied with appetitive conditioning tasks (cf. Ehrenfreund & Allen, 1964).

Another advantage of aversive conditioning is its convenience for studying the effect of age on retention. From postweanling infancy to adulthood, aversion (and detection) thresholds for some shock stimuli do not differ, thus bypassing the horrendous difficulty of controlling deprivation–depletion and incentive variables across age groups (Campbell, 1967).

One additional advantage relates more specifically to our particular aversive-conditioning task. It concerns the necessity for equating degree of learning prior to the retention interval. This can be most readily assured if the task to be retained is learned at the same rate and the final performance level attained does not differ among experimental conditions. By adjusting parameters throughout a series of pilot experiments we have obtained this desirable set of circumstances. Learning of the task to be retained does not differ for animals which have or have not had prior conflicting learning, if the following procedure is used.

The apparatus involved in these experiments consists simply of two compartments, one white and one black, separated by a door (see Fig. 1). A

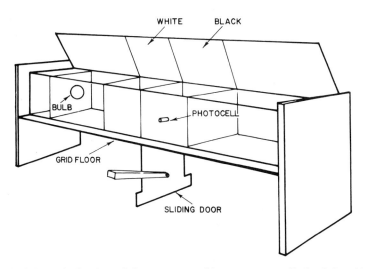

Fig. 1. Schematic drawing of the one-way avoidance apparatus. Each of the white and black compartments measures 10½ x 1⅝ x 5 inches and a ½-inch hurdle separates the compartments. The most significant photocell is pictured, and it functions to turn off the CS during active avoidance, to turn on the foot shock in the black compartment during passive avoidance, and to control the timer.

training series given one set of rats is designed to produce a lapse in retention of active-avoidance learning because of PI. These rats are placed in the white compartment, the door dividing the two compartments is opened, a flashing light comes on, and when the rat steps into the black compartment he is shocked with 1.6 mA for 3 seconds. Most rats enter the black compartment only on the first trial, though some also enter it on one subsequent trial. This passive-avoidance training ends when the rat fails to enter the black compartment within 60 seconds on two consecutive trials. Thirty seconds later the rat is again placed in the white compartment, the door is opened, and the flashing light comes on exactly as before. Five seconds later, however, the rat receives a 1.6 mA shock in the white side and escapes to the black side where, this time, no shock occurs. Soon the rat learns to avoid the white side by crossing over from it to the black before the shock comes on. When this is accomplished on three consecutive trials, at which time crossover latencies drop to about 2 seconds, we say the active avoidance is learned. The intertrial interval throughout all this training is 30 seconds.

After learning active avoidance, the rat is returned to its home cage. If the rat is returned 3 minutes later and given a test trial exactly as in training except that no shock is given anywhere, performance does not differ from that of an animal trained only on the active avoidance. In both conditions, latency to move from white to black usually is less than 5 seconds. This seems to be true 15 minutes later and maybe 30 minutes later as well. However, 60 minutes later a clear lapse of memory occurs for nearly all rats which initially had been given passive-avoidance training: They show considerably less tendency to display the last thing they had learned—to cross over rapidly from white to black—than do those animals that did not receive the initial training.

To summarize, the basic strategy of the experiments which test PI of aversive conditioning included three steps directly analogous to those of the appetitive-conditioning experiments previously mentioned: First, rats were trained sequentially on two opposing response tendencies, initially no-go, then go. Second, the rats' initial responses were measured following various retention intervals. Third, the extent to which the potential source of PI resulted in a memory lapse was determined by comparing the initial responses of these rats with those of rats trained only on the second response tendency.

2. General Results

The increase in the retention lapse between 3 and 60 minutes is seen in Fig. 2a in terms of mean latency to cross over from white to the black side. This function is derived from a preliminary set of our data based upon a composite of various conditions of prior learning.

a. Response Distributions and Their Implications. A clearer picture of the actual effect may be seen in terms of response distributions (see Fig. 2b). Three minutes after active-avoidance training about 80–90% of the scores in both groups qualify as active avoidances. Sixty minutes later this does not change much for the group without the source of PI. However, after this same 60-minute retention interval, the response latencies by rats with prior training on the passive avoidance become more evenly split between those appropriate for active avoidance (5 seconds) and passive avoidance (60 seconds).

Although it is not completely conclusive, this bimodality seems to imply a retrieval effect with both response-tendency attributes potentially available at the retention test. After 3 minutes, the "go" tendency is retrieved most often, and after 60 minutes the joint probability of go and no-go remains about the same, i.e., most latencies are less than 5 seconds (active avoidance) or are the maximum possible 60 seconds (passive avoidance); but after 60 minutes, in contrast to the 3-minute interval, an equal or greater number of responses are no-go rather than go. Without prior passive-avoidance (no-go) training, this does not happen. Thus this interference-induced retention lapse does not seem to conform to a Gestalt-like concept such as "crowding" (cf. Ceraso, 1967) in which the rats given prior passive-avoidance training might be expected to regress toward an average of go and no-go with increasing length of the retention interval. Rather the effect seems best characterized as a transient "mixup" about what to do. According to the present view, the "mixup" increases as the retention interval lengthens because of the corresponding increase in the proportion of cues available for retrieving prior-learning, instead of latest-learning, attributes of memory.

b. Transience of This Case of PI. The transient nature of this retention failure is shown by our consistent failure to obtain similar effects in terms of trials to relearn, even under conditions identical to those yielding big effects of prior learning on the nonshock test trials. After the first relearning trial in which S escapes from foot shock in the white compartment, i.e., upon presentation of cues indicating that the original reinforcement conditions and contingencies are reinstated, the effect seems to be gone. We have diligently pursued instances in which variables influencing PI of retention might be effective in terms of relearning rates by training S to a relearning criterion after the nonshock test trials, but this measure has never yielded a reliable effect of prior learning.

We also have tested this transience more directly. Two groups of rats were given passive-avoidance training followed immediately by active-avoidance training, as previously described. Both groups were given a retention test 24 hours later. The retention test differed for the two groups but only in terms of the consequences of S's first failure to actively avoid. For nearly all rats this first failure to leave the white compartment within 5 seconds occurred on the first test trial. For one group the consequence of such a failure was a footshock given in the white compartment exactly as had occurred during active-avoidance training; this was followed by five nonshock test trials. The effect of this single,

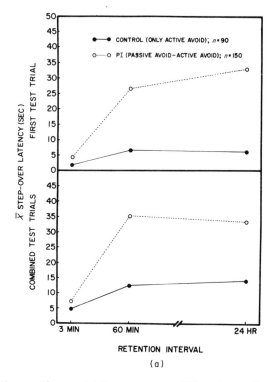

(a)

simulated active-avoidance trial was to nullify the influence of prior passive-avoidance learning for the subsequent test trials: Median response latencies for these rats were less than 1 second on each of the succeeding five test trials which included only two failures to "avoid" among 35 opportunities. In contrast, rats which did not receive footshock on their first failure to "avoid" had median crossover latencies between 59 and 60 seconds on each of the five succeeding test trials, which included 28 failures to "avoid" among 35 opportunities.

Thus, in contrast to the potent effects obtained before reinforcement conditions and contingencies are reinstated, measures taken subsequently, such as number of trials required to relearn, have shown little or no influence exerted by PI variables. This is so, even though the latter measure regularly yields reliable differences caused by retention interval and other pertinent retention variables, some of which are described in later sections.

C. DETERMINANTS OF THE LAPSE IN RETENTION OF ACTIVE AVOIDANCE

In analyzing the influence of prior learning on retention lapse, we have been concerned with two basic effects. The first issue concerns amount of forgetting:

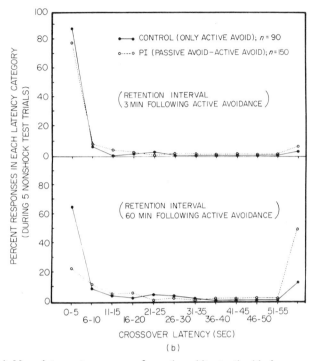

Fig. 2.(a) Mean latency to crossover from the white to the black compartment after different retention intervals following active-avoidance training for *S*s with or without previous passive-avoidance training. This figure combines data from three early experiments, including conditions of prior passive avoidance having varying amounts of effectiveness; the top portion of this figure shows latency on the first (nonshock) test trial and the lower half of the figure shows mean latency combining all five (nonshock) test trials. (b) Frequency distributions showing the percentage of responses in each of the possible latency categories. These are the same data represented in (a).

What determines the asymptotic extent of the lapse? In other words, at the point at which increasing retention intervals no longer affect the rat's retention behavior, what variables determine the extent of the behavioral deviation from most recent learning? The second issue concerns rate of forgetting: What variables determine the time course of the changes in *S*'s tendency to perform in accord with the latest learning?

1. Determinants of the Magnitude of Retention Lapse

In the present case, rats are first trained never to go from white to black, and then are trained to go from white to black in less than 5 seconds. Given a retention interval long enough to produce the greatest lapse, what other variables determine the extent of the rat's failure to go rapidly from white to black?

a. Degree of Prior Training: Number of Reinforced Responses. A parameter that could well control the subsequent latency to enter the black compartment seemed to be the number of aversive stimuli the rat previously had been given there; separate groups of Ss were therefore given 0, 1, 3, or 9 shocks paired with S's presence in the black compartment. Also, for reasons that will become clear shortly, it was expected that the manner in which the rat had encountered the black side and associated aversive circumstances would affect his retention behavior; so half of the rats in each shock condition were placed in the black compartment and given unavoidable shocks while the remaining Ss were placed in the white compartment and permitted to enter the black side on their own.

If the rat was to go from the white to the black side as many as nine times, it was necessary for the shock in the latter to provide only mild punishment (0.5 seconds of 0.16 mA). Since the white side of the apparatus included a flashing light, it was not particularly surprising that the (albino) rats most often entered the preferred darker compartment even though shocked there.

Following passive-avoidance training, all rats learned active avoidance to a criterion of three consecutive avoidances. These variations in prior learning had two effects which were particularly important. First, during passive-avoidance training, the greater the number of aversive experiences with the black side, the more slowly the rat entered it. Second, although this prior learning had no effect 3 minutes after active-avoidance learning, the retention lapse which occurred 60 minutes later was greater the higher the number of prior aversive experiences with the black side. Moreover, it may be concluded that this retention effect was restricted to those rats which had been given their initial experiences with the black side in the same way as their subsequent active avoidance and test trials, i.e., by being placed in the white side and permitted to walk or run to the black side. The corresponding numerical differences (see Fig. 3) among rats given prior unavoidable shocks on the black side were not statistically significant among themselves, nor in relation to the control conditions. The significant interaction between number and type of passive-avoidance trials further supported the difference between training procedures.

Why should a variation in prior training affect the subsequent retention lapse to a greater extent when the prior passive-avoidance trials were conventional "run" trials than when the rats were merely placed in the black side for the unavoidable shocks? Two alternative sources of this effect emphasize the degree of similarity between the memory attributes associated with prior and latest learning and the stimuli noticed at the retention test. First, perhaps the greater similarity between the conventional passive-avoidance trials and active-avoidance learning facilitated the subsequent mixup—retention lapse—about the alternative consequences of go and no-go responding in this situation. Alternatively, the critical aspect of prior learning may have been the similarity between the associated training conditions ("run" vs. "placed") and the circumstances of the retention test ("run"). According to the present view, greater similarity in the

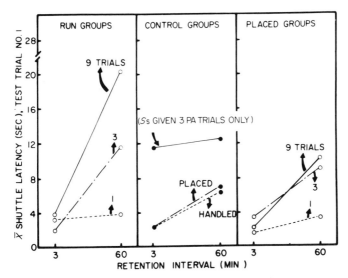

Fig. 3. Mean latency to crossover from the white to the black compartment on the first (nonshock) test trial, either 3 or 60 minutes after the final active-avoidance trial. Active-avoidance training had been preceded by one, three, or nine passive-avoidance trials ("Run" groups and "Placed" groups; first and third panels). In the case of the control groups, the "Placed" condition had active-avoidance training preceded by three placements in the black compartment without shock being delivered, and the "Handled" control had its active-avoidance treatment preceded by three "trials" in which they were only handled and never placed in the apparatus. The two uppermost points represented in the middle panel indicate mean crossover latencies for Ss—included to assess possible "incubation of fear" associated with passive avoidance (PA)—given only three passive-avoidance ("run") trials, no active-avoidance training, and tested after 3 or 60 minutes.

latter sense increases the probability that the retention test will be accompanied by retrieval cues more capable of arousing the memory attributes of prior learning. Since this prior learning is contrary to latest learning, a retention lapse of latest learning becomes more probable after 60 minutes, although a sufficient number of supporting attributes are available after a very short retention interval to prevent a lapse on the 3-minute test. With these alternatives in mind, we may consider a second experiment in which the degree of passive-avoidance training was manipulated in a different way.

   b. *Degree of Prior Training: Number of Passive Avoidance Opportunities.* Passive-avoidance training in this next experiment was given in accord with the "standard" conditions described above: The rat was placed in the white compartment exactly as in active-avoidance training, and when it entered the black compartment, a floor shock of 1.6 mA was delivered for 3 seconds. The rats in this experiment differed in terms of the number of prior-learning opportunities to enter the black compartment (and be shocked there) and in terms of the length of the retention interval between subsequent active-avoidance learning and the retention test. Since most rats enter the black compartment only once under these passive-avoidance training conditions, prior learning experience differed primarily in terms of the number of trials during which the rat was placed in the white compartment, suppressed the tendency to enter black (but, incidentally, did not simply "freeze" on all occasions), and was removed after 60 seconds. Separate groups of rats were given either no passive-avoidance trials prior to active-avoidance learning, one passive-avoidance trial (all Ss in this condition received a shock on this trial, usually entering the black side after about 8 seconds), passive-avoidance training to two successive 60-second suppressions (which usually took 3 or 4 trials in all), or passive-avoidance training to two successive suppressions plus 10 additional trials (which usually took 13 or 14 trials). Finally, this experiment was completed once with adult rats (80–120 days old) and once with very young rats (23–26 days old). Fundamentally the same retention effects occurred for both weanling and adult rats so their data are discussed without further reference to the age variables.

This was the only experiment in this series in which initial rate of learning the active-avoidance task was influenced by prior treatment—active avoidance was learned more rapidly by the adult rats as a direct function of the number of prior passive-avoidance trials. This should not detract from the eventual conclusion that greater lapse in retention of active avoidance is caused by greater numbers of prior passive-avoidance trials, because more rapid attainment of a criterion by itself probably implies more learning on the criterion trial and so better retention. However, these transfer data are of sufficient interest in their own right to merit comment.

The direct relationship between rate of active-avoidance learning and the number of prior passive-avoidance trials may appear curious initially, but compatible results have been reported elsewhere (Bresnahan & Riccio, 1970). Moreover, this relationship seems easily understood by either of two explanations: (1) Perhaps fear of the black compartment decreased with increasing exposure during successful avoidances (cf. Kamin, Brimer, & Black, 1963); or (2) perhaps the additional passive-avoidance trials provided an opportunity for acquisition of nonspecific response dispositions which facilitated transfer to the active avoidance, in opposition to the negative effects of the specific transfer components. Some of the more obvious components of nonspecific transfer, which may be grouped generally under the headings of "warm-up" or "learning-to-learn," probably include decreased resistance to handling, increased habituation of irrelevant stimuli, and improved discrimination of the alternative compartments in the apparatus.

The acquisition data provided no basis for choosing between these alternative explanations, but the retention performance implied that the second might be better. We shall see that after a retention interval of sufficient length, latency to enter the black compartment was directly related to number of prior passive-avoidance trials. One might infer from this that fear of the black compartment was actually increased rather than decreased with more passive-avoidance trials. Alternative implications will not be discussed here since we must move on to the more pertinent retention data.

The results of the retention test are shown in Fig. 4 and may be described

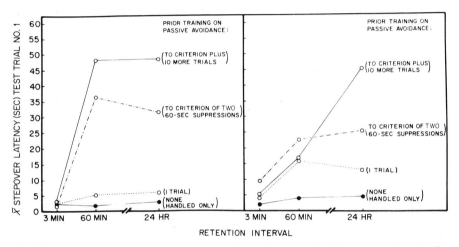

Fig. 4. Mean latency to crossover from white to black on the first (nonshock) test trial following active-avoidance learning by various retention intervals. Active-avoidance learning had been preceded by differing degrees of passive-avoidance training as indicated in the figure. The data shown on the left panel were taken from rats 23–26 days old and those on the right panel from rats 80–100 days old.

simply. After the 3-minute retention interval, performance did not depend on degree of prior learning. However, after 60 minutes and also after 24 hours it did: The higher the degree of prior learning in terms of passive-avoidance opportunities, the greater the rat's retention lapse.

We may ask which memory attributes associated with prior learning were most responsible for the subsequent lapse in retention of active avoidance. Was the retention lapse primarily concerned with what the rat did last or with the most recent consequences of this behavior? The results of these first two experiments suggest that the former alternative is more nearly correct. Variation in number of consequences (shocks) experienced during prior learning affected retention in the first experiment only when the behavior preceding them was similar to that during active avoidance and retention test. The second experiment indicated that even though the number of consequences experienced in prior learning was relatively constant across conditions—means of 1.0, 1.7 and

2.0 shocks were delivered to young and old Ss (combined) in the groups given 1, about 3 or about 13 prior-learning trials—the number of suppression responses emitted during prior learning had an important influence on subsequent retention of the tendency to avoid actively.

It is of interest for the present emphasis on retrieval that the conditions accompanying these prior-learning suppressions, while similar to the conditions of the retention test, were relatively different from those accompanying active-avoidance training. For example, Ss were not shocked during the retention test if they remained in the white compartment longer than 5 seconds so test trials were not immediately preceded by shock experience during the retention-test conditions; this was similar to the most effective passive-avoidance training conditions but different from all but the last few trials of active-avoidance training.

These are the sorts of retrieval cues to which the present point of view would appeal for explaining S's failure to exhibit what it had learned last—stimulus events which are present both during the retention test and also had accompanied prior training to form attributes of the memory for prior learning. As the retention interval increases, certain external and internal stimuli associated with latest learning become progressively less available as retrieval cues while those cues appropriate for retrieving the memory for prior learning become at least relatively more available. For example, home cage events immediately preceding passive-avoidance training—including perhaps eating or resting—will more probably precede the retention test as well if Ss recover for an hour or a day in the relative safety of their home cage than if they have been permitted only 3 minutes rest there after the emotionally disturbing active-avoidance trials. Together with possible temporal changes in inhibitory processes activated during latest learning, these sorts of factors are presumed to account for the increasing interaction of conflicting memories as the retention interval lengthens.

c. Direct Manipulation of Extra-Task ("Contextual") Cues. Although the indirect evidence cited above is not inconsistent with the retrieval framework and fits reasonably well within it, alternative explanations fit equally well. A more direct test is desirable, and this was sought in the next experiment.

The basic feature of this study included presentation during the retention test of an extra-task stimulus ("contextual" cue) which also had been present during prior passive-avoidance training, but which was not present during active-avoidance training. Thus the paradigm provides a correlation between extra-task events and specific tasks, not unlike the "switching" paradigm employed by Asratian (1965).

Specifically, a tone was present during passive-avoidance learning from the time the animal was brought into the room until the end of the last passive-avoidance trial. The tone was turned off during active-avoidance learning and, of course, the rat was not exposed to it while in his home cage during the retention interval. The tone was turned on again during the retention test, being present from the time the rat was brought into the experimental room until after the last test trial. We refer to this tone as a contextual cue because it is not correlated with reinforcement. In a sense, the task could be considered a conditional discrimination if the alternative compartments of the apparatus are viewed as discriminative stimuli. It is undeniably surprising that a tone should acquire stimulus control under these conditions since it was experienced only once, albeit for a long duration. Nevertheless, we had found in pilot experiments that retention lapse for latest learning tended to occur more strongly when the tone was presented during prior learning, not during latest learning, but again during the retention test.

The 28 independent conditions of this experiment differed only in terms of the experimental treatment prior to active-avoidance learning and length of the retention interval interpolated between active-avoidance learning and the retention test. Prior to learning active avoidance, half the rats were given passive-avoidance training to a criterion of two successive, 60-second suppressions, and the other half received equivalent handling but no experimental training. Within each of these conditions, one half of the rats were exposed to tone prior to active avoidance, i.e., during passive-avoidance learning or handling, but the remaining rats were not. For all rats the tone was never present during active-avoidance training but was always present during the retention test. Rate of learning and terminal performance for active avoidance did not differ across conditions, and retention tests were given after either 3 minutes, 15 minutes, 30 minutes, 1 hour, 4 hours, 24 hours, or 21 days.

Presence of the contextual cue at the retention test increased the magnitude of retention lapse attributable to prior learning (see Fig. 5). The index of PI in Fig. 5 is the difference between mean crossover latencies for Ss given and those not given passive-avoidance training before learning active avoidance. The higher the number, the greater the retention lapse of latest learning attributable to PI.

Performance appropriate to latest learning again was impervious to previous learning for a short time, apparently over retention intervals up to 30 minutes. But PI became apparent beginning about an hour after the final active-avoidance trial. Thereafter, PI was most detrimental to retention of active avoidance for those rats which had received their prior training in the presence of the tone, which was also present during the retention test. According to the present interpretation, the retention-test tone functioned to facilitate retrieval of other attributes of the prior-learning memory (e.g., suppression of the tendency to run to black) increasing the probability of behavior associated with that memory.

The above experiment, though including 28 separate conditions, had only five rats per cell. Perhaps because of this, the interaction between retention interval and presence of the contextual cue had only borderline statistical significance.

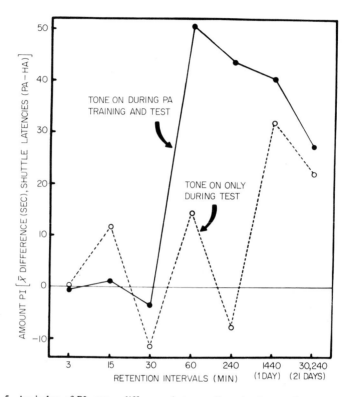

Fig. 5. An index of PI—mean difference between the retention performance (crossover latency on first test trial) of rats given (PA) or not given (HA) prior passive-avoidance training—is shown as a function of retention interval. The solid line represents data from *S*s presented with a contextual tone during their experience preceding active avoidance, not during active-avoidance training, and again during the retention test. Rats whose performance is represented by the dotted line were presented with the tone only during the retention test.

However, this particular result was important for our theoretical framework, and it was necessary to resolve immediately any doubts about the validity of the phenomenon. Therefore two subsequent experiments of a similar nature were completed. One used a tone as above for the contextual cue, the other employed a combination of cues including the experimenter's glove, brighter ambient lighting, and a low frequency buzz from a motor. The details of these experiments need not be mentioned except to note that the results were in agreement with the first experiment.

In our view, the importance of this effect is not limited to the increased influence of passive avoidance on retention produced when a contextual cue is present during both passive-avoidance training and the retention test. The presence of contextual stimuli which are irrelevant to but associated with original acquisition of instrumental behavior often has been shown to enhance subsequent performance of that behavior (e.g., Donahoe, McCroskery, & Richardson, 1970). We have shown in addition that the effect is contingent upon the length of the retention interval between training and test: Presentation of a contextual cue associated with prior conflicting learning did not influence retention shortly after learning, but it did facilitate the interference effect of this prior learning on a later retention test.

There are at least two possible interpretations of this effect within the present theoretical framework. (1) Perhaps it is generally true that rate of forgetting is inversely related to the number of redundant attributes associated with the target memory. We assume that as the retention interval lengthens, certain attributes are lost, and others may become less discriminable from attributes of competing memories. Thus, although functionally redundant at the short retention interval, an attribute such as the contextual tone may effectively function at a later time to retrieve the passive-avoidance memory, i.e., as the retention interval lengthens and other attributes are lost or become less effective in retrieval. This alternative can involve the assumption that the probability of retrieving attributes associated with active avoidance remains constant over the retention intervals. (2) An alternative interpretation might assume maintenance of a constant retrieval probability for passive-avoidance attributes. In so doing, the above effect could be based upon the decrement in the probability of retrieving active-avoidance attributes as the retention interval increases. The contextual cue was not present during active-avoidance learning but was during the retention test. One could argue that while this did not influence retrieval of active-avoidance attributes at the shorter retention intervals—at which time a relatively large number of supporting attributes for active avoidance were available—retrieval was impaired by this stimulus change after longer intervals. As a consequence, passive-avoidance attributes were more probably retrieved after the longer intervals and passive-avoidance behavior exhibited to a greater extent.

The second alternative seems the less likely. It has been shown that, under some circumstances, retention performance *improves* as the retention interval increases, if a major stimulus change occurs between original learning and the retention test (e.g., Desiderato, Butler, & Meyer, 1966; McAllister & McAllister, 1968; see also Gleitman, Chapter 1). Also our data indicated that Ss trained only on the active-avoidance task showed little decrease in performance between 3-minute and 60-minute retention tests, even though the contextual tone was present during the test but not during original learning.

d. Summary. Tentatively, then, there appear to be at least two factors determining the magnitude of the retention lapse. The first is the strength of the conflicting response tendency established during prior learning, defined in terms of the number of times it was practiced effectively. The second is the number of stimulus elements (potential memory attributes) noticed by S during the retention test which also had been noticed by S during prior, but not latest, learning. Presumably these elements selectively act to retrieve the memory attributes of prior learning.

Why do these characteristics of prior learning—as well as the basic PI effect—have no influence on retention behavior in this situation until 30-60 minutes have elapsed since latest learning? This important question concerning rate of forgetting will be considered next.

2. Determinants of the Rate of Forgetting: When is the Retention Lapse Most Pronounced?

The experiments listed above indicate that variations in conditions of prior learning, though quite effective after 60 minutes, do not influence latest learning after shorter retention intervals. The 3-minute retention interval consistently has shown this property—actually, it is the only interval less than one hour which has been examined extensively—although the third experiment above suggests that this period of "perfect retention" may extend for as long as 30 minutes. Even so, 3 minutes is six times as long as the intertrial interval, clearly qualifying it as a legitimate retention interval. The point is that even when given an appropriate retention test, *after a short retention interval the rat does what he last learned to do regardless of prior experience.* What mechanisms control the length of the interval after which effects of prior learning may be observed? There appear to be no clearly applicable data available at the moment. We have considered two approaches to this problem, the first fundamentally independent of the passage of time, the second temporally dependent.

a. Stimulus-Change Explanation. The most parsimonious approach would apply the same retrieval mechanisms that determine the magnitude of the retention lapse. For example, one effective retrieval cue or memory attribute appropriate for retrieval of the tendency to passively avoid could be sequential experiences (cf. Capaldi, 1967). Immediately prior to passive-avoidance training, the rat simply has been stationed in his home cage doing his usual "thing" since the beginning of the working day. Perhaps as the subsequent retention interval spent in his home cage exceeds 3 minutes and approaches 60 minutes, his experience just prior to being returned to the experimental room more closely approximates that preceding his passive-avoidance training. It should be recalled that active-avoidance training began only 30 seconds after the conclusion of passive-avoidance training, thus including antecedent events quite different from those associated with passive avoidance.

b. Inhibition-Dissipation Explanation. Perhaps general response tendencies such as "go" and "no-go" have reciprocally inhibiting mechanisms. Subsequent to learning passive avoidance, active-avoidance performance may require the rat to inhibit the entire response system responsible for the no-go tendency. During the development of efficient active-avoidance performance, the rat may concurrently develop and maintain an active inhibition of the no-go tendency. The inhibition may subsequently be removed either by spontaneous dissipation or some disinhibitory stimulus, the probability of either of the latter increasing the longer the retention interval. Perhaps it is worth noting that such a mechanism is not dissimilar from a recent application (Postman, Stark, & Fraser, 1968) of the concept of "generalized competition" (e.g., Postman, 1961). Postman *et al.* convincingly argue that this mechanism of inhibition can account for many temporal changes of interference effects on verbal retention. It is probably only coincidental that the similar temporal changes studied by Postman *et al.* occurred over retention intervals similar in length to those reported here.

c. Preliminary Study of the Determinants of Rate of Forgetting. We have completed three sets of experiments which, although not directly designed to decide between the above two theoretical alternatives, have nevertheless tended to confirm our expectations concerning the parameters which control the length of the maximally effective retention interval. Conclusions up to this point may be stated briefly and somewhat tentatively because this series is not yet complete.

Less influence of prior learning is found after a 60-minute interval the greater the amount and distribution of practice on the most recently learned task. The reduction in PI observed in a training-test interval of one hour does not, however, reflect a general decrease in the magnitude of the influence of prior learning, but represents its postponement to a later time, e.g., until a 24-hour interval has passed. Specifically, as the criterion for active-avoidance learning is increased from three consecutive avoidances to three consecutive avoidances plus an additional 10 trials, or three consecutive avoidances plus an additional 20 trials, crossover latencies become relatively shorter after a 1-hour retention interval; by comparison, performance after 3 minutes or 24 hours is affected to a lesser extent. Similarly, as the interval between successive active-avoidance trials is increased from 30 seconds to 12 minutes or 24 hours, crossover latencies for those *S*s given prior passive-avoidance training become relatively shorter after a 1-hour retention interval.

The interval separating passive-avoidance and active-avoidance training seems to have a similar effect on retention. As this interval is increased from 30 seconds to 60 minutes or 24 hours, the influence of prior passive-avoidance training becomes less evident at the 60-minute retention test. We have found a similar decrease in the influence of prior passive-avoidance training when separated from active-avoidance training by 21 days although this is at least partially attributable to a decrease in retention of the passive-avoidance learning itself. Increasing the interval between prior learning and latest learning also has been reported by Underwood and Freund (1968) to decrease PI in human memory.

These results seem equally compatible with either of the theoretical alternatives cited above. More generally, however, the effects of increasing degree of latest learning, the interval between prior and latest learning, and the intertrial interval of latest learning appear to reflect the operation of retrieval rather than decay processes of memory. It may not be unreasonable to expect that rate of memory "decay" might depend upon degree of learning (e.g., on the basis of attribute redundancy), but the other main effects apparently would require several additional, complicating assumptions for a proper explanation based on "simple" decay of memory. Moreover, a decay-based explanation of the interaction between these variables and presence of prior learning would appear to become further encumbered by still more assumptions which are not needed for a retrieval-based explanation.

Whatever the most suitable interpretation, the effect of prior learning on retention of go, no-go behavior in rats appears to be general and robust. But this has not been found for retention of discrimination learning, a fact which we must now consider.

D. Temporal Changes in Interference with Retention of Discrimination Learning

Choice behavior has great appeal as a fundamental response measure for comparative psychology. Its convenience for evaluating species or age effects is obvious, and so it has seemed to us that this would be the logical place to begin for a comparative study of interference and forgetting. In this respect data available up to the present time have been discouraging. Indeed, a comparative analysis of interference and forgetting in terms of choice behavior by the rat (or pigeon) has been precluded by the elementary failure to obtain consistently a significant increase in the rate of forgetting choice behavior given a potential source of PI. A summary of this work, including some rather optimistic suggestions concerning why most experiments had failed to obtain the effect, has been presented elsewhere (Spear, 1967). So this section will concentrate upon a few of the more recent successes along with some of a long series of largely negative results from our laboratory.

1. Studies Concerning PI in Retention of Discrimination Learning

Surprisingly little experimental attention has been given to the manner in which conflicting memories of prior and later discrimination learning interact during a retention interval. Consequently, there are few data suitable to explain the interactions which have been found or why they often do not appear as expected. Often experiments nominally directed toward this problem have used inappropriate methods (e.g., have included only a single retention interval), and only rarely have they included operations of much analytical value.

a. Failure to Obtain PI in Retention of Simple Position Discrimination. At present, six papers have appeared showing poorer retention of choice behavior as a consequence of PI.[2] It may be important that all of these have involved difficult discrimination tasks, i.e., difficult relative to the simplest discrimination of position in which no dimensions are irrelevant, the correct response (of two) is continuously reinforced, and the incorrect response is never reinforced. The latter task usually is well learned within 30 trials depending upon such things as reinforcer magnitude and saliency of cues correlated with position. Retention of this simple position discrimination has never been found to be significantly impaired by prior learning of the reverse discrimination. It is, of course, impossible to know how many experiments of this sort have obtained negative results. As one example, however, approximately 10 experimental variations on this theme are recorded among our notebooks of unpublished data, none of which show sufficiently consistent interactions between presence of a PI source and retention interval to warrant mention in print. The author is aware of similar sets of negative results obtained in other laboratories as well.

Surely it is unlikely that prior learning has absolutely no influence on retention even considering only rats as subjects and this simplest type of position discrimination. The possible combinations of pertinent parameter values, even within these restrictions, must remain infinitely larger in number than those tested so far. Indeed, although the resistance to PI that the rat demonstrates under these conditions remains astounding, some relatively subtle experimental variations of these conditions have resulted in substantial PI.

b. Some Evidence for PI in Retention of Position Discrimination. Two published experiments have found poorer retention of modified versions of the position discrimination due to previous acquisition of the reverse discrimination (Gleitman & Jung, 1963; Koppenaal & Jagoda, 1968). Gleitman and Jung used a variation of a fixed-ratio schedule for reinforcement of the correct response; this study is described more completely in Chapter 1 by Gleitman. Koppenaal and Jagoda (1968) modified their reversal task so that the incorrect alternative during the retention test had never previously been incorrect. The latter was accomplished by providing that a straightahead alternative at the choice point be incorrect for both learning tasks while the correct alternative always was a 90° turn, right for the first task and left for the second, or vice versa. Thus, as the first task, continuous reinforcement was provided for turning, say right (or, for some Ss, left) in contrast to no reinforcement for simply running straight ahead. The second task had left "correct" and straight ahead again as "incorrect." The retention test had left correct, right incorrect. Unfortunately for our purpose, it

---

[2] Other useful experiments (e.g., Chiszar & Spear, 1969, Exp. 1; Cole & Hopkins, 1968; Gonzalez *et al.,* 1967) have shown that prior learning influences subsequent performance, perhaps reflecting impairment of retention processes; but these experiments, in one way or another, do not meet the present definition of retention studies.

is impossible to evaluate whether either of the procedural variations introduced in the Gleitman-Jung and Koppenaal-Jagoda experiments actually increased the PI effect or, if so, which aspect of the procedural variations was responsible for the effect: neither experiment included conditions showing that PI would have been less effective if the position discrimination tasks had been more conventional.

It is perhaps worth mentioning that we have been unable to replicate the results of Koppenaal and Jagoda while repeating their basic design in a series of four experiments. However, among other differences which may be noted *ad hoc,* learning was more rapid in our experiments, perhaps because the start and goal boxes were not interchanged as in Koppenaal and Jagoda's automated apparatus, and final degree of learning apparently was lower. This should not detract from Koppenaal and Jagoda's convincing and useful results; the replication failure is noted in order to emphasize the fragile nature of the influence of PI on retention of choice behavior.

To our knowledge, the only remaining experiments showing an influence of PI on retention have used visual discrimination tasks. Three of these have included discrete-trial bar-pressing (Maier, Allaway, & Gleitman, 1967; Maier & Gleitman, 1967; Spitzner & Spear, 1967) and the fourth used a T-maze brightness discrimination (Chiszar & Spear, 1968b). The study by Maier *et al.* (1967) is perhaps the most analytical of these.

The learning to be retained by rats in the most critical experimental condition of the Maier *et al.* (1967) experiment, as in the Koppenaal-Jagoda study, did not involve nonreinforcement of the originally correct alternative. However, Maier *et al.* found that the detrimental influence of PI on retention was no stronger under this condition than when nonreinforcement of the originally correct alternative *was* involved in the task prior to the retention interval.

c. Spontaneous Recovery as a Special Case of PI. The results of Maier *et al.* provide indirect support for our contention that the influences of PI on retention, or more generally, the interactions of conflicting memories over time, constitute pervasive phenomena of animal learning. Our theoretical framework has suggested that similar retention processes underlie the effects observed following successively learned conflicting responses or consequences of responses, including instances involving acquisition vs. extinction, small reinforcers vs. large reinforcers, active avoidance vs. passive avoidance, and acquisition vs. reversal of a discrimination. Hence our view implies that extinction is not a unique process, contrary to some theories, and similarly, spontaneous recovery is not somehow different from similar changes in "retention" found subsequent to counterconditioning.

Maier *et al.* (1967) found that PI in retention was about equal whether or not latest learning involved extinction of an originally correct response. Furthermore, a subsequent study by Crowder, Cole, and Boucher (1968) agreed in finding no greater "recovery" of an originally correct choice over a 48-hour interval subsequent to extinction than subsequent to counterconditioning. Thus the results of both studies, together with those of Koppenaal and Jagoda (1968), are consistent with the present view that spontaneous recovery from extinction is fundamentally a special case of PI in retention, and if inhibitory mechanisms are involved in spontaneous recovery, they are no different from those concerned with the storage and retrieval of any conflicting memories.

2. *Factors Opposing PI with Choice Behavior*

A number of hypotheses have been suggested by investigators to explain why they did or did not find PI in retention of discrimination learning. Most of these explanations, including some of the author's, have leaned heavily on analogies with theories of PI developed in the area of verbal learning (e.g., Koppenaal & Jagoda, 1968; Spear, 1967). [Various stages of development of these verbal-learning theories may be followed in reviews by Keppel (1968), Postman (1961), and Spear (1970).] However, we believe, as does Winograd (see Chapter 7), that it now may be more profitable to base such explanations on factors intrinsic to the problem presented an animal which is given a retention test under circumstances nominally appropriate to either of two incompatible responses. In our view, the animal must derive "instructions" from the balance of retrieval cues most closely associated with one of the responses.

It follows from this view that if PI is to occur, the retention test must be given in such a way as to minimize retrieval cues appropriate to latest learning in favor of those of prior learning. Surely the latest reinforcement conditions and contingencies qualify as such retrieval cues, but these must be reinstated if the retention test is given as relearning of the latest task. On this basis the influence of PI should be inherently transient with a relearning test. We have mentioned above that such transiency has been found among all, or nearly all, studies of go, no-go behavior, and this also holds to a lesser degree for choice-behavior experiments (e.g., Chiszar & Spear, 1968b; but see Maier & Gleitman, 1967). An effect of PI rarely is found after the recall trial, and if it is observed in terms of relearning rate, it dissipates rapidly.

The transience of PI presents measurement problems which are particularly evident when S's choice between two alternatives is the dependent variable. Choice prior to reinstatement of reinforcement conditions and contingencies— i.e., choice on the first retention trial—must be more sensitive than subsequent choices to the influence of PI. But without extraordinary behavioral control, the combination of typical variance in behavior between perfect retention and purely random choosing plus the single binary measure yields unacceptable statistical power without massive numbers of subjects. One alternative is to give a number of trials at the retention test before introducing any reinforcement contingencies at all. An extinction test in which no reinforcer is available in either alternative appears to meet the latter condition (except for the continued absence of a reinforcer associated with the most recently nonreinforced alternative). If so, we should expect greater influence of PI with an extinction test than when the retention test is relearning. That the results have not conformed to our expectations may be seen in two brief examples.

   a. Effect of the Retention Test on Retention after a Reversal Shift. In the first example—an experiment by J. Spitzner and the author in which training on a brightness discrimination prior to a reversal shift served as the source of PI—the basic results indicated that this PI source *did* impair retention.

   The discrimination task in this experiment required the rat to press either of two levers depending upon the state of a light placed above each lever. Half of the rats received one food pellet when they pressed the bar under the light that was turned on, but not if they pressed the bar under the light that was turned off; and the remaining half had the reverse contingencies. The opportunity to solve this discrimination was presented on a discrete-trial basis with presentation of a trial contingent upon the rat's touching a handle on the wall of the operant chamber opposite the discriminanda.

   All rats were allowed 25 trials per day. After considerable initial training in which only a single bar was present, rats in the PI condition were trained on their original brightness discrimination until they attained the criterion of at least 80% correct responses on two consecutive days; Ss in the C (control) condition continued practicing on a single lever during this time. Twenty-four hours after reaching their criterion on the original discrimination, Ss in the PI group were given the reverse discrimination to learn to the same criterion; at about the same time, Ss in the C condition learned their sole discrimination task to the same criterion. Retention tests then were given after intervals of either 1 day, 14 days, or 28 days. For half of the rats the retention test consisted of reacquiring the discrimination most recently learned; for the other half the retention test was extinction. However, except for the difference in reinforcement contingencies, the retention tests were conducted with precisely the same details of procedure.

   First, consider only those rats given relearning of the most recent task as the retention test. Retention scores are plotted in Fig. 6 in terms of the difference between the percentages of correct responses on the last day of acquisition and the first day of the retention test. From these data we may conclude that PI from prior learning of the reverse discrimination *did* interfere with subsequent retention, and with greater effect the longer the retention interval. These results are in close agreement with those of Maier and Gleitman (1967). This replication is not particularly surprising because the methods of the two experiments were quite similar. However, our prediction that the PI effect would be stronger for those rats given extinction as their retention test was not verified. If anything, the trend is in the opposite direction (see Fig. 6), although the mysterious

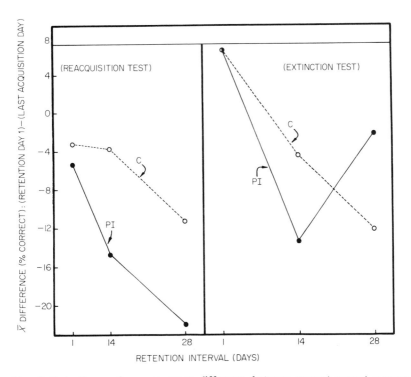

Fig. 6. Retention performance—mean difference between percent correct responses during first day of the retention test and last day of acquisition—is shown for the C and PI conditions. The Ss in the C condition had learned only the task being tested, and PI Ss had learned this task after prior training on the reverse brightness discrimination. The data in the left panel were obtained from rats given reacquisition as the retention test, and those in the right panel from Ss given extinction as the retention test.

reversal in the effect for Ss given extinction after a 28-day interval probably should not be taken too seriously at this time.

  b. *Effect of the Retention Test on Retention after a Nonreversal Shift.* The above prediction—that a greater PI effect should be found if the retention test consists of extinction rather than reacquisition—subsequently was tested in another experiment completed by J. Spitzner and the author. Although generally similar to the preceding experiment, the procedure and results of the present study differed in one important respect each: The PI condition included a nonreversal rather than a reversal shift, and PI with retention was not observed.

  Basically the design of this experiment included two PI conditions in which rats were shifted from either a brightness to a position discrimination or vice versa prior to the retention interval. Two control conditions also were included in which rats learned only a position or only a brightness discrimination prior to the retention interval. The same retention intervals and types of retention test were presented as in the preceding experiment.

The results shown in Fig. 7 require little comment: PI had no consistent influence on retention whether retention was measured in terms of relearning or extinction.

Again, we had expected that PI with retention of choice behavior would

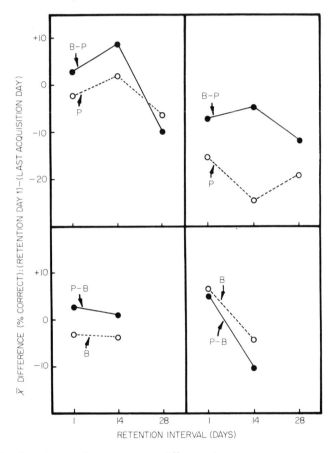

Fig. 7. Retention performance—mean difference in percent correct responses during the first day of the retention test and last day of acquisition—is shown for rats given and not given prior training on a conflicting discrimination. Rats given prior training had a nonreversal shift prior to the retention interval, "P" refers to position discrimination and "B" refers to brightness. Rats given only a single discrimination task (P or B) prior to the retention interval received an amount of prior training on a single neutral bar approximately equivalent to that of Ss given prior discrimination training. The panels on the left represent rats given reacquisition as their retention tests, those on the right represent rats given extinction as their retention test.

occur more strongly when the retention test consisted of extinction than when relearning was the test. During relearning, presentation of the reinforcement conditions consistent with the most recently correct alternative was assumed to arouse critical memory attributes appropriate for a clearer distinction between the conflicting memories of prior and latest learning. Alternatively, if extinction were given as the retention test it seemed that the reinforcement conditions and contingencies would be neutral with respect to prior and latest learning, so the conflicting memories would be less easily reconciled and PI would result. The source of the failure to verify these expectations is not clear. It may simply lie in our reasoning within the present theoretical framework. Or the problem may rest in our lack of understanding supplementary aspects of rat behavior such as the cue value of reward in discrimination learning (cf. D'Amato & Jagoda, 1960; Hill, Cotton, & Clayton, 1963) and effects of extinction given after different retention intervals with and without prior reversal or nonreversal shifts.

Thus the problem of measuring PI with discrimination learning remains unresolved. Perhaps the best solution is to obtain a number of retention measures on the same subject as has been accomplished when the influence of PI on successive reversals has been studied (e.g., Gonzalez, Behrend, & Bitterman, 1967; Mackintosh, McGonigle, Holgate, & Vanderver, 1968). Unfortunately, many experimental questions do not permit the application of this solution.

3. An Auxiliary Effect of Prior Learning on Retention of Discrimination Learning

It is notable that PI may affect some retention behavior without directly influencing the percentage of responses appropriate to the most recent discrimination. Reference here is to "intrusions," in this case the probability that an error on a retention test would have been correct during prior learning. This probability can be assessed independently of retention of the latest learning if a nonreversal shift is given. For example, suppose prior learning reinforces responses to light vs. no-light irrespective of position, and latest learning reinforces left vs. right irrespective of the light. During the subsequent retention test we may count the number of responses given at the incorrect position when the associated light is on compared to the number given there when the light is off. The calculation of "intrusions" in this way is shown in the lower half of Fig. 8 based upon the data from the nonreversal-shift experiment cited above. Clearly, prior learning affects retention behavior in this respect considerably more than would be expected by chance. Moreover, in agreement with the more general effect of PI on retention, there is a tendency for these intrusions to increase in probability with increases in length of the retention interval. In contrast, the upper panel of this figure repeats the result described above: that the prior learning was not at all detrimental to retention in terms of S's absolute

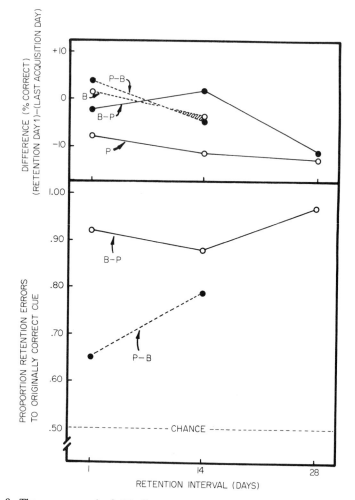

Fig. 8. The upper panel of this figure repeats the data shown in Fig. 7, this time combining the two kinds of retention tests (reacquisition and extinction). The negligible suggestion of forgetting increased by PI is shown by the crosshatched lines. The dependent variable in the lower panel is an index of the probability of an intrusion error after various retention intervals following the acquisition of the second discrimination (retention test given on second task).

probability of giving the most recently acquired response. Kehoe (1963) found similar results on a 5-key discrimination problem with pigeons as subjects.

We had initiated this experiment because a nonreversal shift presumably permits less extinction of prior learning than a reversal shift. Our reasons for expecting that PI would appear more strongly if prior learning were less deeply extinguished corresponded to those of Koppenaal and Jagoda (1968) although the operations for reducing extinction were, of course, quite different. The success of our procedure was ambiguous. In spite of the unexpected absence of PI with retention of the most recently correct response, prior learning did influence retention behavior in terms of errors. Again, retention errors were more likely to be committed when the currently incorrect stimulus was paired with a stimulus which had been correct during prior learning than when paired with a previously incorrect stimulus. How may this be reconciled with the fact that, overall, the probability of an error during the retention test was no greater for rats with prior learning than without? There appear to be at least three plausible explanations.

First, perhaps preexperimental, idiosyncratic response biases contribute as much interference in the control groups as do the "response biases" experimentally induced by prior learning in the PI groups. If so, and if these preexperimental response biases could be accurately identified, we should find the same sort of intrusions in the retention behavior of the control Ss. That is, retention errors made by the control Ss also should be found to be more probable when the incorrect stimulus is accompanied by the stimulus preferred due to preexperimental response bias.

Second, perhaps conservation processes of retention operate in such a way that the rat quickly reacts to and recovers from momentary susceptibility to interference; and otherwise, PI affects behavior only given a temporary disposition to make an error anyway.

The third explanation is based upon maintenance of S's attention to the relevant stimulus dimension.[3] Suppose that errors committed on the retention test are due to a momentary failure to attend to the dimension which most recently was relevant. Then rats given prior learning involving a different stimulus dimension should be more likely to attend to that dimension on a retention error and should respond accordingly. On the other hand, Ss without such prior learning should respond randomly if they fail to attend to the most recently relevant dimension during the retention test. The underlying assumption, of course, is that the probability of attending to the most recently relevant dimension is independent of, or at least not inversely related to, the number of other relevant dimensions to which S has attended in previous experimental encounters (cf. Honig, 1969).

In any case it does appear that, as in verbal learning (cf. Keppel, 1968), interference-induced "overt intrusions" may have little relationship to interference-induced "forgetting" in the rat.

4. Conclusions

It seems best to withhold further speculation concerning the conditions which may be responsible for PI in retention of discrimination learning.

[3] The fundamental idea for this explanation was suggested by Werner K. Honig.

Generally, however, we believe that retrieval processes probably will be found to determine these conditions, although this belief so far is based on only indirect evidence. Two experiments by Chiszar and Spear (1969) have provided some of this evidence.

In both studies, rats learned a simple position discrimination in a T-maze and then learned the reverse discrimination in either the same T-maze or a similar one located in a different room. The consequent influence of a few orthogonal variables involving retention interval was determined. The specific details need not be repeated here, although the results become more interesting when it is realized that initial transfer from one maze to the other was essentially perfect, i.e., nearly all rats made the turning response they had just learned when given their first trial in a different T-maze.

One generalization which emerges from this study is particularly relevant here. Overall, the results indicated that changes in behavior which otherwise occur with longer retention intervals become muted when stimuli accompanying reveral learning differ from those which accompanied acquisition. In other words, conflicting memories of discrimination learning appear less likely to interact with the effects of a retention interval and with retrieval cues at the retention test when their respective attributes are relatively dissociated. Thus questions surrounding interference and retention of discrimination learning may be most readily resolved through consideration of the similarity between the attributes of the potentially conflicting memories along with the proportion of retrieval cues at the retention test which correspond more closely with the memory attributes of prior learning than with those of latest learning.

IV. Retention Loss over Intermediate Intervals

The final set of experiments to be discussed in this chapter concerns retention loss over intervals of intermediate length (e.g., Kamin, 1957). Specifically, the "Kamin Effect" is said to occur when retention of aversive conditioning is reasonably good immediately and 24 hours later, but poor when tested after intermediate retention intervals of, say, 1 or 4 hours.

It should be clear, however, that the definition of this nonmonotonic retention function is not restricted to these specific intervals. There are some cases in which similar functions have been defined by intervals of quite different magnitude (cf. Suboski, Marquis, Black, & Platenius, 1970). Furthermore, to anticipate the discussion below, there have been a variety of instances showing that retention increases monotonically with increasing length of retention interval (e.g., Spevack & Suboski, 1969) which may be viewed as consequences of the same processes causing the effect originally measured by Kamin. Therefore we shall use the term "Kamin Effect" to refer to phenomena which may be defined by operations a good deal different from those employed by Kamin (1957).

A. An Example: Kamin Effect as a Function of Age

Although the Kamin Effect is not a new phenomenon it has continued to arouse empirical and theoretical interest, and we believe it has sufficient generality to require that it be dealt with in any serious consideration of animal forgetting. An example of this effect is shown in Fig. 9. These data, some of which have appeared in a paper by Klein and Spear (1969), show the Kamin Effect as a function of age. Initially, only the panel on the far right will be discussed, data from rats aged 25–26 days. The dotted line represents the number of trials required to relearn the one-way active-avoidance task (see Fig. 9) to a criterion of five consecutive avoidances following original training to a single avoidance.

It is clear that the inverted U-shaped function, the Kamin Effect, did occur for 25- to 26-day-old rats under these conditions: Relearning was poorer after intermediate retention intervals than after shorter or longer intervals. Although not shown on this figure, this basic effect was found in another experiment to be no different from that obtained in rats aged 90–100 days (Klein & Spear, 1969). This is consistent with the general conclusion that rats aged about 25 days and older do not differ in original learning of this task nor in its short-term retention (i.e., intervals of 24 hours or less, cf. Campbell, 1967). However, as can be seen in the other panels in Fig. 9, the inverted U-shaped function was less likely to

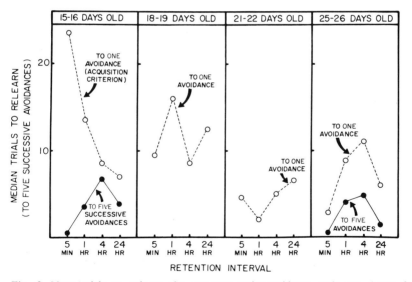

Fig. 9. Mean trials to relearn the one-way active-avoidance task are shown for independent groups of weanling and preweanling rats after each of four retention intervals following original training. The dotted line represents data of rats originally trained until they performed a single avoidance, and the solid line represents data from rats originally trained to a criterion of five consecutive avoidances.

occur among rats younger than 25 days, at least when they previously were trained to a single avoidance.

This result seemed consistent with certain hormonal explanations in which the Kamin Effect was presumed dependent upon the existence of adrenal corticosteroids (Brush & Levine, 1966). Levine and Mullins (1968) have reported data indicating that 15-day-old rats emit fewer of these corticosteroids when stressed than 25-day-old rats, and 25-day-old rats apparently do not differ from older rats in this respect. However, since rate of learning this task decreases below 25 days, and since learning may occur on the criterion trial, attainment of the criterion of one avoidance by a 15-day-old rat may represent less learning than attainment of the same nominal criterion by a 25-day-old rat (cf. Underwood, 1964). Thus, age may have been confounded with degree of original learning—another variable which affects forgetting in terms of loss (i.e., trials to relearn). This possibility caused us to proceed with tests involving higher degrees of learning before directly attributing the age-related differences in the Kamin Effect to age-related differences in corticosteroid output.

The panel on the far right of Fig. 9 shows that the Kamin Effect also occurred when 25-day-old rats had originally learned the active avoidance through five consecutive avoidances. In fact, the statistical significance of the effect was greater after the higher degree of learning, perhaps because of a reduction in variability. The more interesting outcome of the increase in degree of learning occurred with the 15- to 16-day-old animals: The function relating relearning performance to retention interval began to approximate a Kamin Effect for these infants, although short of statistical significance. This result compelled us to do one more experiment with rats younger than 25 days, increasing the degree of learning still further.

Rats between the ages of 15 and 18 days were given original learning to 10 consecutive avoidances. Fortunately, almost all of these Ss which reached a criterion of five consecutive avoidances also attained 10 consecutive avoidances, thus eliminating the possibility of a confound by subject selection in comparison with the previous experiments. The results shown in Fig. 10 clearly indicate that the Kamin Effect can occur in 15- to 18-day-old rats given a sufficient degree of original learning.

B. THEORETICAL ANALYSIS AND INITIAL TEST

The above series of experiments (Klein & Spear, 1969) cannot really be said to have eliminated any existing explanation of the Kamin Effect. But it did create some doubt in our minds about these explanations, particularly the fear-based and hormonal theories, which otherwise had seemed reasonably convincing. This in turn gave us sufficient courage to proceed with an alternative explanation emphasizing the availability of attribute-related retrieval cues as the fundamental determinant of the Kamin Effect.

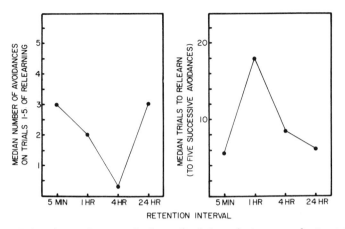

Fig. 10. Relearning performance is shown for independent groups of rats aged 15–18 days. Relearning followed original training to 10 consecutive avoidances by one of four intervals.

1. Explanations of the Kamin Effect

Until recently, there had been about as many different explanations of the Kamin Effect as relevant experiments. However, no single explanation had followed from a more general theory, nor generated much research outside of the basic effect itself.

The basic feature of the Kamin Effect is a persistent failure to behave in accord with latest learning after intervals a great deal shorter than are usually required to produce similar retention failure. The adjective "persistent" emphasizes the fact that this is not merely a transient mixup about what to do. In fact, our data have indicated that the Kamin Effect often is measurable in terms of trials to criterion even though it is not apparent in terms of behavior on the first few retention trials, whether shock is delivered at the retention test or not (see also Kamin, 1963). Thus, in our terms, the phenomenon involves loss rather than lapse. Still, it does appear to be a matter of retrieval rather than storage failure, because otherwise retention would be expected to be no better after 24 hours than after 4 hours. Of course, it may be argued that it was the motivation—not the memorial predisposition—that was absent at the intermediate retention intervals. Indeed, this kind of argument seems to be the most prevalent of the explanations of the Kamin Effect which, in turn, may be divided into two categories.

a. Theories Implicating Temporally Dependent Changes in "Fear." The central process in the first category of explanations is "incubation of fear," a hypothesized increase in the intensity of this emotional state over a retention interval of relative inactivity. This intuitively appealing process seems to be rather sensitive to procedural variation—we have found no evidence for it among several attempts in our laboratory, for example—and there is some doubt as to its validity (McAllister & McAllister, 1967). However, others have placed such serious theoretical stock in it that it demands consideration (e.g., Denny & Ditchman, 1962; Spevack & Suboski, 1969).

The typical, and reasonable, assumption used in these explanations is that the efficiency of acquiring or performing certain avoidance tasks is a correlate of fear intensity and so follows the same time course. For some theorists and some tasks performance has been presumed poorer with more intense fear; in other cases performance has been presumed to improve with increasing fear. Accordingly, the theoretical application of "incubation of fear" to the Kamin Effect depends upon the hypothesized time course of the former. In some cases it has been employed primarily to account for the decrease in performance between the immediate and intermediate tests, and in others to explain the increase in performance between the intermediate and later tests. Additionally, some second process necessarily has been invoked to account for that aspect of the nonmonotonic course of the Kamin Effect which is not explained by incubation of fear.

b. Theories Implicating Changes in Performance-Controlling Hormones. The second explanatory category has been based upon the physiological changes which accompany apparent changes in emotional state and follow a reasonably regular time course subsequent to stress. One specific aspect of these changes, variation in the number of circulating adrenal corticosteroids, has been emphasized by Brush and his co-workers (e.g., Brush & Levine, 1966). They have argued that avoidance behavior will be effective to the extent that these circulating steroids are present. If the rat is returned to its home cage after original training, the plasma steroid level decreases rapidly and reaches a low point about an hour later, when avoidance is correspondingly poor. This accounts for half of the curvilinear relationship; for the remainder, these investigators have suggested that the improved performance after 24 hours may be attributable to increased sensitivity of the adrenal system.

c. A Retrieval Interpretation. The point of view taken in this chapter suggests a different kind of explanation, one emphasizing retrieval of the predisposition to behave in accord with latest learning. In our view, relearning is poorer after 1 or 4 hours than 5 minutes or 24 hours after original learning because cues for retrieving the original memory are less likely to be present during the 1- or 4-hour tests.

The ultimate task is to identify those critical retrieval cues that change or disappear after intermediate retention intervals. It is possible that external stimulus changes play an important role in this, e.g., changes in the ambient noise level or temperature along with other aspects of the rat's circadian cycle related to the time of day at which training and testing are given. Perhaps a more likely source might be the large number of internal physiological changes which take place during and after stress, at a maximum in some cases following onset or offset of stress by intermediate intervals and restabilizing, say, 24 hours later (e.g., Selye, 1946). One such change might be the modification of steroid levels mentioned above. I must quickly admit that the final three sets of experiments reported here do not primarily involve identification of the hypothetical attributes. Their results simply tend to favor an interpretation of the Kamin Effect in terms of memory retrieval rather than the alternative explanations outlined above.

2. Initial Test of the Retrieval Explanation: The Negative Transfer Paradigm

The rationale of the following two experiments may be stated briefly: Instead of testing retention in terms of *relearning* of aversive conditioning acquired 5 minutes, 1 hour, 4 hours, or 24 hours earlier, retention was tested in terms of *transfer* to a conflicting aversive-conditioning task. If the Kamin Effect is attributable to temporal changes in fear or physiological correlates of fear which maximize after intermediate intervals, and if these changes result in an animal correspondingly less capable of handling the circumstances of aversive conditioning, the transfer task should be learned most efficiently after very short or very long retention intervals and most poorly after intermediate intervals. In other words, the "fear-incubation" and "hormonal-change" explanations predict equivalent results whether the retention test involves positive transfer (i.e., relearning in this case) or negative transfer. However, if the Kamin Effect is a consequence of the relative ineffectiveness of memory retrieval 1 or 4 hours after original learning, the negative-transfer effect of original learning should be less after these intermediate intervals. Therefore, the retrieval hypothesis predicts poorest learning of the new task 5 minutes or 24 hours after original learning and best performance 1 or 4 hours later. Somewhat to my surprise this is precisely what was found.

a. General Procedure. Both of these experiments (Klein & Spear, 1970a) were conducted in our laboratory and employed the active-avoidance and passive-avoidance tasks already described. Because 25-day-old rats were used as Ss (preliminary data suggested these younger rats were more susceptible to interference), a scaled-down version of the adult-sized apparatus was used. Also a few parameters of the aversive stimulus were changed to maximize negative-transfer effects.

b. Learn Active Avoidance, Test Passive Avoidance. In the first experiment, rats were given active-avoidance training to five consecutive avoidances. Separate groups then were trained on the conflicting passive-avoidance task either 5 minutes, 1 hour, 4 hours, or 24 hours later. For active avoidance, 1.6 mA served as the shock intensity, while 0.2 mA for a 1-second duration served as the aversive stimulus in the case of passive avoidance. Passive-avoidance training was taken to a criterion of 3 successive suppressions of 60 seconds each.

The trial-by-trial changes in crossover latency during passive-avoidance training are shown in Fig. 11. The lower two lines reflect performance after 5 minutes or 24 hours. It is clear that a great deal of negative transfer occurred in these conditions relative to the performance of naive subjects. In contrast, the prior active-avoidance training had almost negligible effect on acquisition of the

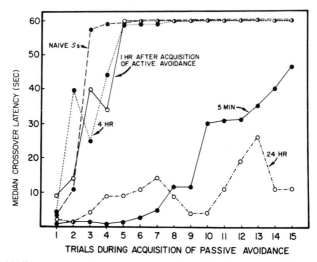

Fig. 11. Median crossover latencies on successive trials of passive-avoidance training are shown for rats which differed in terms of the retention interval by which this passive-avoidance training followed active-avoidance learning in the same one-way avoidance apparatus. The curve designated "naive" shows the acquisition performance of Ss given identical passive-avoidance training but without previous active-avoidance learning.

passive avoidance for Ss given the 1- or 4-hour retention interval: The increases in response latencies by these Ss were scarcely distinguishable from those of naive subjects. These same results are reflected in terms of trials to criterion on the passive-avoidance task (see Fig. 12).

c. *Learn Passive Avoidance, Test Active Avoidance.* Since these results also could be expected if the rats became exhausted, immobile or otherwise debilitated 1 or 4 hours after original training, the second experiment reversed the order of the tasks. All rats were given passive-avoidance training as before, followed after 5 minutes, 1 hour, 4 hours, or 24 hours by active-avoidance training. The results in terms of crossover latencies are shown in Fig. 13. The relationships in terms of running speed are exactly opposite those of the previous experiment, but because of the change in order of the tasks, they reflect the same fact: Efficiency of acquiring the second task was considerably greater 1 or 4 hours after having learned the conflicting task than 5 minutes or

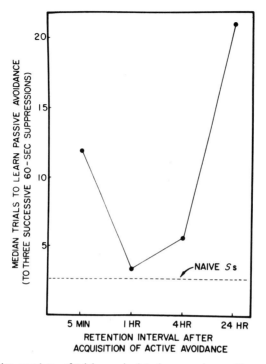

Fig. 12. Median numbers of trials required to learn passive avoidance to a criterion of three consecutive 60-second suppressions are shown as a function of retention interval following active-avoidance learning.

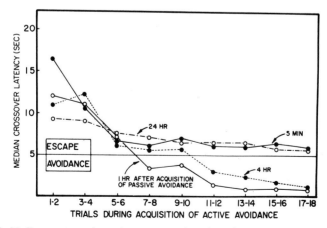

Fig. 13. Median crossover latencies on successive trials of the one-way active-avoidance task are shown for rats after one of four retention intervals following passive-avoidance learning.

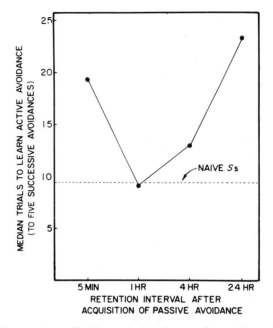

Fig. 14. Median numbers of trials required to learn active avoidance to a criterion of three consecutive avoidances are shown as a function of retention interval following passive-avoidance learning. Acquisition performance for naive Ss—Ss given active-avoidance training without prior passive avoidance—is represented by the dotted lines.

24 hours later. Similarly, the criterion of active-avoidance learning was attained more rapidly by rats tested 1 or 4 hours later than by rats tested after 5 minutes or 24 hours (see Fig. 14).

d. Interpretation and Comparison with Other Evidence. Taken together, these two experiments argue against an interpretation in terms of temporal changes in fear or temporal changes in the mobility of the animal. Rather, they appear consistent with the conclusion that the Kamin Effect occurs because fewer attribute-arousing cues are available to effect retrieval of the target memory attribute of original learning 1 or 4 hours after aversive conditioning than immediately or 24 hours later.

Two comparable studies have appeared in the literature, one by Baum (1968), the other by Bintz (1970). The results of both studies agree with the present interpretation. We have found no others with which to compare our results.

Baum (1968) trained rats on active avoidance with a procedure very similar to ours but to a more stringent criterion of original learning (to 10 consecutive avoidances). Either 10 seconds, 1 hour, 4 hours, 22 hours, or 44 hours after original learning, Baum trained these rats on the "reverse" avoidance task. This task also was one-way active avoidance but given in the direction opposite that of acquisition: The rat was placed in the previously "safe" compartment where it received foot shock 5 seconds after the door opened unless it entered the previously "dangerous" compartment where foot shock no longer was delivered. This particular reversal procedure, in which the response remains the same as in acquisition but certain stimuli are reversed, is known to result in net *positive* transfer (Baum, 1965).

According to our interpretation whenever original training impairs subsequent avoidance learning—i.e., whenever negative transfer occurs—learning will be best after intermediate retention intervals; but if original training facilitates subsequent avoidance learning—i.e., positive transfer—a conventional Kamin Effect is likely to govern the degree of this facilitation. For example, Brush, Myer & Palmer (1963) found a Kamin Effect when avoidance training followed signalled-escape training (i.e., fear conditioning with a 0.5-second CS–UCS interval and escape possible) by various intervals, but the interval by which avoidance training followed unsignalled escape training had no effect on avoidance behavior. On this basis our interpretation predicts, and the data from Brush *et al.* seems to agree, that the prior signalled-escape training had a greater positive transfer effect on avoidance learning than the prior unsignalled-escape training, and the latter probably had neither positive nor negative transfer effects. Thus, in our terms, failure to retrieve the memory of original training after intermediate retention intervals results in relative impairment of subsequent avoidance learning only after fear conditioning. This reasoning implies that Baum (1968) should have obtained a conventional Kamin Effect with his (positive transfer) procedure, and he did.

The experiments by Bintz (1970) are so complex and thorough that justice cannot be done to them in a few sentences here. The reader may obtain details of the study in the original article. Briefly, Bintz included a negative transfer condition and found, in agreement with our results, that performance on the transfer task was better after an intermediate retention interval following original learning than after shorter or longer intervals.

C. Facilitation of Retrieval When Stress Precedes Retention Test: Reactivation of Memory Attributes

The retrieval cues presumed to be modified or absent after intermediate retention intervals constitute a hypothetical construct at this point in our research. However, it has seemed likely that the most potent cues of this sort, especially for simple avoidance behavior, are the specific chemical, neurological and, perhaps, muscular states of the organism, i.e., internal cues.

We have referred above to the variety of physiological systems aroused by stress which subside during subsequent retention intervals of relative inactivity by *S*. Certain of these appear to be less readily reactivated after an intermediate interval than after longer intervals (e.g., Levine & Brush, 1967). There are several other internal changes induced by stress which reach maximum after an intermediate interval and may serve as the cause of retrieval failure, although neither specific endogenous processes nor their most important mode of influence in effecting changes in internal stimuli can be asserted with confidence at this time. For example, we have mentioned the work showing a close correlation between concentration of plasma corticosteroids and retention performance following avoidance learning (Brush & Levine, 1966; King, 1969; Levine & Brush, 1967). It may be that detection of plasma corticosteroids provides a salient internal cue state for the rat; it may be that the most potent stimulus effect is not directly due to the corticosteroids but to their inhibitory influence on ACTH activity (e.g., Weiss, McEwen, Silva, & Kalkut, 1968); or perhaps corticosteroids provide their most potent influence as stimuli only after reaching their binding sites in the brain.

Thus the specific nature of the changes in the retrieval cues which are most evident after intermediate retention intervals are not specified, nor do we believe it is necessary to do so at this time. Nevertheless, perhaps the internal state of the organism at the time of original learning may be recaptured after an intermediate retention interval if a stress source similar to that of original learning is presented to *S*s just prior to the retention test. To the extent that this is accomplished, relearning performance should be no different after intermediate retention intervals than after shorter or longer intervals.

1. General Procedure

Three experiments were completed by S. Klein and the author to investigate this problem. Each involved separate groups of adult rats tested for retention of an avoidance task after short (10 minutes), intermediate (2½ hours), or long (24 hours) intervals. At each retention interval the retention test for one group of rats was immediately preceded by presentation of five inescapable shocks delivered in an apparatus unlike the avoidance apparatus; the other group received a retention test as usual without prior shock. Separate control groups received only the inescapable shocks to evaluate the influence of such shocks on avoidance performance in the absence of prior training.

The three parallel experiments differed in terms of the avoidance task used for original learning and the retention test. In one experiment, active avoidance was learned originally and relearning was given as the retention test; in another, active avoidance was learned originally and passive-avoidance learning served as the retention test; and in the third, passive avoidance was learned originally and active-avoidance learning was the retention test.

2. Results

The specific results of these experiments have been presented elsewhere

(Klein & Spear, 1970b). We shall briefly discuss the experiment in which original learning was active avoidance and retention test was passive avoidance. Each experiment yielded the same conclusions, though this one provides somewhat more continuity with the studies discussed in the next section.

The results of the retention test are shown in terms of the number of trials required to learn passive avoidance subsequent to one of several treatments (see Fig. 15). Rats in the basic experimental condition, the "Shock" group, together

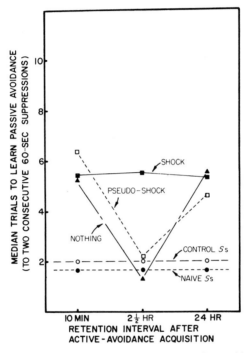

Fig. 15. Median numbers of trials required to learn passive avoidance are shown as a function of retention interval after one-way active-avoidance learning ("Shock," "Pseudo-Shock," and "Nothing" conditions) or related control treatments.

with those in the "Nothing" and "Psuedo-shock" conditions had learned active avoidance to a criterion of five consecutive avoidances prior to the retention interval.

Rats in the Shock group were returned to the experimental room 7½ minutes prior to the end of their retention interval and given five inescapable shocks (1.6 mA, for 1 second each) with a 30-second interval between shocks. In accord with our prediction, retention performance by Ss in the shock condition was not different after an intermediate interval than after shorter or longer intervals. In contrast, Ss treated exactly as the Shock Ss except for the actual foot shock (Pseudo-shock condition) and Ss given no particular treatment between original learning and the retention test (Nothing condition) learned passive avoidance in fewer trials after an intermediate interval than after shorter or longer intervals, thus reflecting poorest retention of the conflicting active avoidance after the intermediate interval.

Moreover, Ss in the latter two groups performed no differently after the intermediate retention interval than either Ss which learned passive avoidance without prior experimental experience ("Naive" Ss) or Ss which also learned only passive avoidance but following the same inescapable shock treatment given in the Shock condition ("Control" Ss). This suggests, as do previous experiments, that few or no memory attributes of original active-avoidance learning are retrieved after an intermediate retention interval.

Finally, comparisons between the experimental conditions suggested that the inescapable shocks preceding the retention test permitted retrieval of the memory of original active-avoidance learning as efficiently after an intermediate retention interval as after shorter or longer intervals: As indicated in Fig. 15, passive-avoidance performance after an intermediate interval in the Shock condition did not differ from performance after shorter or longer intervals in either the Pseudo-shock or Nothing conditions.

3. Interpretation

We believe these results are consistent with the present view of forgetting as retrieval failure. According to this view, the poorer retention of avoidance learning found after intermediate retention intervals is due to the absence of retrieval cues needed to arouse memory attributes of original learning; but if conditions approximating those of original learning are reinstated at the time of the retention test these memory attributes may be aroused as effectively, and memory retrieval made as probable, as ordinarily is found after shorter or longer retention intervals.

It may be noted that these results are contrary to expectations derived from the assumption that rats ordinarily show poorer retention performance after an intermediate interval only because they are limited in their capacity to cope with aversive conditioning. Presentation of shocks prior to testing must have the partial effect of "priming" the depleted adrenal system and thus, if there is a motivational effect, it should increase the rat's capacity to cope. However, the present results with the negative-transfer paradigm would indicate that rats cope more effectively after intermediate retention intervals *unless* shocks precede the retention test.

D. AFTEREFFECTS OF STRESS AS MEMORY ATTRIBUTES[4]

We have concluded that the Kamin Effect is caused by poor memory retrieval after intermediate retention intervals. In turn, it has appeared that this retrieval

[4] Most of the data described in this section have appeared in a paper by Spear, Klein & Riley (1971).

failure is at least partially due to greater change in the internal state of the organism after an intermediate retention interval compared to shorter or longer intervals. We believe that the internal states of the organism may serve as memory attributes and retrieval cues. In the present case, the variation in internal state is presumably caused by the particular physiological condition of the rat after intermediate-length intervals following stress, which differs from its condition after shorter or longer intervals.

The similarity between this interpretation and explanations of "state-dependent learning" has been noted above and now becomes more explicit. The paradigm for state-dependent learning has involved training the organism while it is either under the influence of drugs or not, followed by a retention test given under either the same drug state which prevailed during learning or the alternative state. If retention performance is relatively poor when testing occurs under the alternative state—i.e., when the internal retrieval cues present during the retention test are not in accord with the corresponding memory attributes of original learning—then "state-dependent learning" is said to have occurred.

1. Inescapable Shock as Stress

We completed an extensive series of experiments, including five complete studies and a number of smaller experiments, intended to test directly the hypothesis that the stress-induced organismic conditions existing after intermediate intervals are sufficiently different from those prevailing 24 hours later to provide circumstances for state-dependent learning. The basic experimental condition in these experiments included initial training of rats on an avoidance-learning task at some time after they had been exposed to inescapable shocks; then 24 hours after original training, at which time all rats were presumed to be about equally recovered from the effects of the original stress, a test for retention of avoidance learning was given. According to the above interpretation of the Kamin Effect, those rats whose original avoidance training followed stress by an intermediate interval should have learned under a physiological state which was more deviant from that following subsequent recovery than was the case for rats whose stress had preceded original avoidance training by shorter or longer intervals. Therefore rats given an intermediate interval between stress and original learning should have shown poorer retention.

Our analyses throughout this series of experiments were consistent in showing that our predictions were not verified. However, our persistent behavior in the face of these consistently negative results was not completely irrational. From the start, we were aware of the following two reasons to expect that our empirical predictions might not be upheld even though the hypothesis might be correct.

(1) The intensity of original stress probably determines not only the magnitude of the physiological aftereffects but also the time course of these aftereffects, i.e., the interval following stress at which the aftereffects are

maximal (cf. Haltmeyer, Denenberg, & Zarrow, 1967). Having no information concerning the time course of the physiological aftereffects caused by our particular inescapable-shock treatment, we may have failed to capture the maximal aftereffects with the particular intervals we employed between inescapable shock and original avoidance learning.

(2) It is known that nominally equivalent amounts of shock given rats which can control their presentation of shock compared to rats which cannot may result in different physiological and behavioral consequences (e.g., Seligman, Maier, & Solomon, 1971). So it did not seem unlikely that different kinds of temporally dependent processes might be set in motion by experience with controllable shock (as in active-avoidance learning) and uncontrollable shock (as in our inescapable-shock treatment). Perhaps the physiological aftereffects most conducive to state-dependent learning are those which occur as a consequence of controllable stress.

2. Avoidance Learning as Stress

Both of the above reasons suggested that we pursue this state-dependent-learning paradigm further, but using active-avoidance learning as the original stress rather than inescapable shock. If our interpretation of the Kamin Effect is correct, training to a criterion of five consecutive avoidances on our active-avoidance task must be sufficient to produce effectively different physiological states after an intermediate interval compared to shorter or longer intervals. Therefore we conducted an experiment in which passive-avoidance learning followed active-avoidance learning by either 5 minutes, 2½ hours or 24 hours. Passive-avoidance relearning then was given 24 hours later. To prevent original overlearning of passive avoidance and thus to permit differences to occur on the retention test, a low criterion of original learning was employed. (A criterion of a single 5-second or a 30-second suppression was employed in separate replications of this experiment; the results of these two replications did not differ and are combined for further consideration.)

When the test for retention of passive avoidance was given 24 hours later, the 2½-hour group had poorer retention than either the 5-minute or 24-hour group, as predicted (see Fig. 16). To check on possible differences in terms of the original degree of passive-avoidance learning, an additional three groups corresponding to those above were tested for retention of passive avoidance after only 2½ hours. If the poor performance of the 2½-hour group given the 24-hour test was due to a lower degree of original learning rather than inferior subsequent retention, the same effect should have appeared after the shorter retention interval, i.e., before all Ss had recovered so completely from original stress. However, this effect clearly did not occur when retention was tested after the shorter interval (see Fig. 16) thus eliminating the possibility of contamination by differing degrees of original learning.

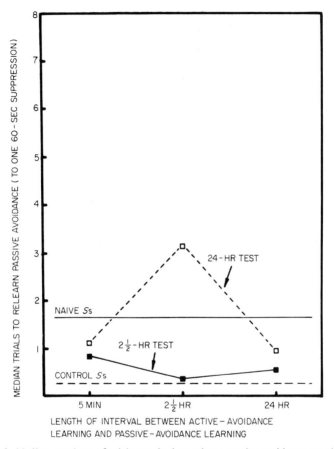

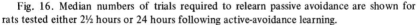

Fig. 16. Median numbers of trials required to relearn passive avoidance are shown for rats tested either 2½ hours or 24 hours following active-avoidance learning.

3. Comparison of Controllable and Uncontrollable Sources of Stress

The next experiment directly compared the relative effectiveness of controllable and uncontrollable shock as agents for subsequent state-dependent learning. Perhaps inescapable shock would have created conditions sufficient for state-dependent learning in our initial experiments if the proper number of shocks had been distributed differently; perhaps active-avoidance learning to five consecutive avoidances "accidentally" arranges for the proper number and distribution of shocks whether controllable or not.

Therefore, rats which could neither avoid nor escape their shocks were "yoked" to rats given active-avoidance learning as initial stress in order to equate the number, duration, and distribution of shocks received. Identical shock was delivered concurrently to both members of a yoked pair, but the experimental member received its shock as a consequence of failure to avoid in the avoidance apparatus, while the control member was shocked in a different, neutral location. Within each shock treatment (i.e., controllable vs. uncontrollable) three groups of rats were included and given either a 5-minute, 2½-hour, or 24-hour interval between shock treatment and passive-avoidance training. A retention test—passive-avoidance relearning—then was given all Ss 24 hours later.

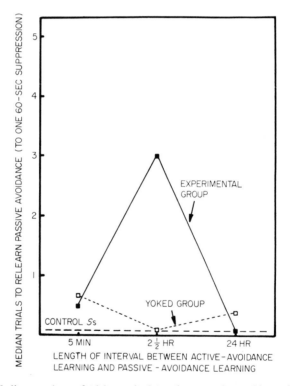

Fig. 17. Median numbers of trials required to relearn passive avoidance 24 hours after original learning are shown for rats initially given active-avoidance training (Experimental Group) compared to those initially given uncontrollable shocks (Yoked Group).

The results of this experiment are shown in Fig. 17. As the earlier experiments had suggested, a state-dependent-learning effect occurred for Ss given an intermediate interval between original stress and original passive-avoidance learning, but only if that original stress had involved active-avoidance learning. Again, such state-dependent learning apparently confirms our memory-retrieval interpretation of the Kamin Effect.

V. General Discussion

A. Comments on Alternative Interpretation: Interaction of Conflicting Memories

We examined the interaction of memories associated with "go" behavior and those associated with "no-go" behavior. Under some cases the go behavior represented approach to a desirable consequence, in other cases it represented avoidance of an undesirable consequence. In some cases the memory associated with go was acquired prior to that associated with no-go, in other cases the opposite order of acquisition occurred.

In spite of these and other variations, there appeared to be a common tendency for conflicting memories of go and no-go behavior to interact over time—the probability of behaving in accord with most recent learning decreased more rapidly given prior acquisition of the conflicting memory than without it However, we found that if such memories were acquired through appetitive conditioning, they were rather mysteriously intractable; and they were therefore inconvenient tools for investigating the general interaction of conflicting go and no-go memories. Instead we concentrated our efforts on a paradigm in which a rat first learns to passively avoid one of two distinct sides of a box and then to actively avoid the other side prior to a retention interval and test.

Rats given this sequence of tasks forget very rapidly: They suffer a sizeable memory "lapse"—i.e., transient failure to perform in accord with latest learning—after only 60 minutes, whereas control Ss trained only on active avoidance show little if any memory lapse after 60 minutes. The magnitude of the lapse produced by the interaction of conflicting memories was found to be directly related to the degree of prior conflicting learning and to the similarity between the circumstances of prior learning and those of the retention test. The length of the retention interval after which the memory lapse maximized—i.e., rate of forgetting—was found to depend upon length of the interval separating acquisition of the conflicting memories and degree of latest learning Although a number of competing interpretations could not be entirely excluded, we concluded that these characteristics of interacting memories for go and no-go behavior are consistent with the present view of forgetting as retrieval failure.

The interaction of conflicting memories associated with discrimination learning was relatively ambiguous and permitted fewer tests of the present retrieval framework. Our work in this area has lagged behind that concerned with go, no-go behavior because we have not found interactions between discrimination-learning memories with similar generality or consistency However, there were some indications that these memories also interact over retention intervals in accord with the present retrieval framework.

Throughout our study of this problem, the fact that has impressed me perhaps more than any other is the apparent capacity of the rat for resisting interaction of conflicting memories. Among a wide variety of circumstances and after quite long retention intervals, the integrity of the rat's memory for most recent learning seems to be unimpaired by previous contradictory learning. Moreover, we have often found no greater forgetting due to experimentally induced interference in spite of considerable forgetting caused by other, unknown factors (e.g., perhaps interaction between the most recently acquired memory and extraexperimental memories). We have noted our frequent frustration in this regard in the course of establishing a viable experimental paradigm for this problem area.

1. Appetitive vs. Aversive Reinforcers

Perhaps it is significant that our attempts to study the interaction of memories associated with go and no-go behavior or discrimination learning have met with little success when appetitive conditioning is involved, although quite robust effects of this sort may be obtained—at least in the case of go and no-go behavior—with aversive conditioning. This is not to deny that our specific appetitive-conditioning tests may have been insensitive for these purposes or that more robust interference effects involving appetitive conditioning could be obtained with other techniques. Indeed Bintz, Braud, and Brown (1970) have recently reported effects interpretable as interactions of memories associated with go and no-go behavior, but using a response measure which may be more sensitive than the running speed measures used in our studies. In addition, Thomas and his colleagues (e.g., Thomas & Burr, 1969) have been measuring the interfering effects of prior discrimination learning (appetitively motivated) on retention of a similar discrimination task in terms of subsequent generalization gradients. This technique may prove quite superior to those previously used to study interference and forgetting of discrimination learning. However, there is reason to believe that our greater success with aversive- than appetitive-conditioning paradigms may be more than an accident of measurement technique.

a. Relative Forgetting Rates of Appetitive and Aversive Conditioning. There appears to be a widespread belief that generalization gradients associated with aversive conditioning (i.e., avoidance learning, escape learning, and classical conditioning with aversive stimuli) not only become flatter with increasing length of the interval between training and testing but do so at a more rapid rate than is the case with appetitive conditioning. This comparison between appetitive and aversive conditioning has been entirely implicit. Indeed, convincing empirical substantiation would be a difficult and lengthy task.

Nevertheless if this belief represents fact, it may explain the differential robustness of the PI found in the go, no-go behaviors controlled by appetitive, compared to aversive, reinforcers. For example, one of the primary determinants of PI is likely the extent to which S discriminates latest from prior learning on the basis of some temporal attributes (also see Gleitman, Chapter 1). It follows that such a discrimination will deteriorate more rapidly over a given retention interval if the conflicting tasks included aversive reinforcers than if they involved appetitive reinforcers.

b. Characteristic Responses to Ambiguity by the Frightened Rat. Proactive interference in retention of appetitive conditioning, when effective, has appeared symmetrically among "go" and "no-go" behaviors, i.e., prior go learning has been shown to interfere with retention of no-go, and also prior no-go learning has been shown to impair retention of go behavior. However, we have not found PI in retention of no-go learning attributable to prior learning of go with our aversive-conditioning paradigm. We have not pursued this problem in depth and have simply attributed it to methodological circumstances, but we may find that this is incorrect. Perhaps when rats have experienced only aversive reinforcers in a given situation, their dominant response to ambiguity is no-go.

Assuming that the conflicting memories of no-go and go interact to a greater extent the longer the retention interval, we may assume that ambiguity concerning the reinforcement contingencies also increases. Thus we may always find an increase in the tendency to no-go as the retention interval increases, given acquisition of go and no-go aversive conditioning in either order. If so, this would form a plausible basis for explaining the "incubation" phenomenon measured following passive-avoidance learning (e.g., Spevack & Suboski, 1969).

c. Conditions Responsible for the Kamin Effect May Facilitate PI. Finally, an interaction between conditions conducive to PI and those favorable to the Kamin Effect may contribute to the advantage of the aversive-conditioning paradigm for studying conflicting memories. This interaction involves the following considerations.

First, note that the Kamin Effect probably is restricted to aversive conditioning. Perhaps this may be extended to include special appetitive circumstances involving a great deal of stress. Nevertheless there have been no published reports showing that simple retention of appetitive conditioning is functionally related to the retention interval in the same way as aversive conditioning. Of course, one cannot know how often retention of appetitive conditioning has been measured under appropriate circumstances without finding a Kamin Effect. Such a negative result would be of little interest and not published. But we have never found a Kamin Effect with appetitive conditioning of go behavior when applicable conditions (usually control conditions to estimate the influence of prior reinforcement conditions on retention of appetitive conditioning) have been used in our laboratory.

Next recall that when a common task was used in our studies, the poorest retention found in the Kamin Effect occurred after about the same interval (about 1 hour) as was required for the appearance of PI. (For the moment we will ignore the fact that PI occurs most often in the form of a lapse and not a loss, while the converse is true for the Kamin Effect.) We have attributed the Kamin Effect to the relative sparsity of appropriate retrieval cues and consequent lack of memory attributes from latest learning aroused after this 1-hour retention interval. It is reasonable to expect that PI might be more effective under these same circumstances, i.e., when the memory attributes of prior learning need compete less with those of latest learning, especially if prior learning were itself so distributed as to be less susceptible to the processes underlying the Kamin Effect.

This interpretation involves a potentially more general principle: Perhaps prior learning is more likely to influence retention the fewer the number of memory attributes aroused which were associated with latest learning. For example, we have found that the retention interval after which PI occurs apparently is longer the higher the degree of latest learning. Probably a higher degree of learning implies the formation of more associated memory attributes (and it may be no more than that). Thus a longer retention interval is required to pare down the number of latest-learning attributes which are aroused at the retention test, if prior learning is to influence retention performance.

B. Comments on Alternative Interpretation: Retention Functions Which Do Not Decrease Monotonically with Retention Intervals

The exaggerated retention failure which follows avoidance learning by intermediate retention intervals has been attributed to exaggerated retrieval failure. The two most convincing interpretations of this phenomenon had appeared to be those postulating changes in "fear" over a retention interval, either increasing or decreasing between original training and completion of an intermediate interval, and those suggesting that, because of hormonal aftereffects of stress, the animal is relatively incapable of coping with an avoidance task after intermediate intervals compared to shorter or longer intervals. Four sets of experiments convinced us that a retrieval explanation might be more appropriate.

First we found that infant rats, which are known to exhibit relatively little hormonal change consequent to stress, do nevertheless show poorer retention of avoidance learning after intermediate intervals compared to shorter or longer intervals. This became evident when differences in learning capacity were taken into account. Next, among rats given a conflicting avoidance task as their retention test—a task in which better memory of original learning results in poorer performance—we found that those tested after an intermediate retention interval avoided *more* effectively than those tested after shorter or longer intervals.

Two additional sets of experiments followed to elaborate our memory-retrieval interpretation. In the first it was shown that retention after intermediate intervals could be made more effective if certain conditions were reinstated at the time of the retention test. The reinstated conditions were those which presumably had been present during original learning, but less so after intermediate intervals than after shorter or longer intervals. In fact, given such reinstatement and presumed reactivation of corresponding memory attributes, memory retrieval was as effective after intermediate intervals as after shorter or longer intervals.

Finally, when a new task was learned following avoidance learning by an intermediate interval, the new memory attributes apparently were sufficiently different—both from those attributes acquired following avoidance learning by shorter or longer intervals and the retrieval cues present 24 hours after the new task was learned—to result in poorer retrieval of the new memory at the 24-hour retention test.

1. Role of Behavioral Changes Which Are Correlated with Retention Changes

It should be clear that the present interpretation of the Kamin Effect does not exclude potential contributions by temporal changes in emotional responses such as "incubation of fear" (e.g., Spevack & Suboski, 1969), temporal changes in hormonal or neurological processes which might influence the animal's capacity to cope with a certain task (e.g., Brush & Levine, 1966), or temporal changes in the cue value of certain odors (e.g., King, 1969). Indeed any or all of these may be critical components determining the availability of retrieval cues capable of arousing the target memory attribute. The present interpretation simply categorizes these factors as contributing indirectly to memory retrieval rather than directly to memory application.

C. THEORETICAL DISTINCTIONS BETWEEN FAILURE IN MEMORY STORAGE AND FAILURE IN MEMORY RETRIEVAL

Compared to the area of human memory, the development of theory and of clear theoretical issues associated with animal memory is immature. However, the tools available to the student of animal memory—relatively greater control which may be exerted over the animal's experiential and genetic history and greater freedom in modifying and measuring the animal's physiological structure—are particularly suited for resolution of the general issue concerning the relative importance of failures in memory storage and failures in memory retrieval. Accordingly, this is the issue within which the theoretical lines are becoming drawn with increasing clarity by the study of animal memory. Three examples may illustrate this issue more concretely.

1. Retrograde Amnesia

Amnesic agents such as electroconvulsive shock, puromycin, and potassium chloride do indeed decrease the probability of behavior appropriate to the most recent learning, if administered shortly after this learning. This has been interpreted most often as reflecting disruption of the storage process, in particular interruption with consolidation of the memory representing this latest learning (e.g., McGaugh, 1966). However, this interpretation has been questioned within the past few years on several bases, primarily because behavior appropriate to latest learning may subsequently occur without additional training, given the appropriate "retrieval" circumstances. This suggests that the memory was adequately stored but the amnesic agent interrupted the process of retrieval. Lewis (1969) and Miller (1969) have described a variety of techniques by which memories apparently destroyed by an amnesic agent subsequently have been shown to be completely or partially intact, at least in terms of the animal's behavior. A more recent example of this may serve as a specific illustration.

Avis and Carlton (1968) trained rats on a CER (conditioned emotional response) task, injected potassium chloride into the hippocampus after 24 hours, and found 3 days later that the memory for CER apparently had been obliterated. However, a subsequent study in Carlton's laboratory employed the same treatment as Avis and Carlton but included longer operation-test intervals (Hughes, 1969). This experiment showed that memory for the CER was in fact retrievable, with recovery of the CER beginning after about 21 days. Moreover, additional evidence from Carlton's laboratory has indicated that this "amnesic" effect may be obtained even though the KCl is injected prior to, or even several days after, training of the CER (Carlton, personal communication, 1969).

The point of this is not to suggest that disruption of memory consolidation cannot be affected by some amnesic agents since something like this certainly seems likely. Rather, we must be cautious in attributing all retrograde amnesia effects to loss of the memory from storage (cf. Lewis & Maher, 1965). It is beginning to appear that only some treatments permanently exclude a memory from storage and sorting these out from treatments that in fact influence only retrieval would appear to be a critical empirical problem in the study of memory.

2. Development of Memory

Another case involving a theoretical contrast between storage and retrieval processes is the greater forgetting occurring in younger organisms. It is now quite clear that although weanling rats may learn with rates and characteristics identical to adults, they forget more over long retention intervals (e.g., Campbell, 1967). The magnitude of this effect is impressive and its importance for understanding behavior hardly can be overestimated.

Specific examples of greater forgetting by younger rats have been reported by Feigley and Spear (1970) using the same active-avoidance and passive-avoidance tasks referred to in our experiments described above. Shock intensity was varied orthogonally to age and retention interval in these experiments. The weanlings and adults did not differ in the number of trials required to attain the original learning criterion of five consecutive active avoidances. Original learning was slightly more rapid with more intense shock, but shock intensity did not enter into any statistically significant effects in terms of the number of trials required to either learn or relearn after an interval. The point of principal interest here is the greater long-term forgetting by the weanlings: Although relearning performance was equal for weanlings and adults after a 1-day retention interval, the weanlings required considerably more trials to relearn than did the adults after 28 days. Retention of passive avoidance yielded similar results: Retention between 1 and 28 days declined considerably more for weanlings than adults.

In view of the immature neurological state of the brain in weanling and preweanling rats, their greater forgetting may be attributed to storage failure, e.g., the memory storage may be less redundant and so more "fragile" in the sense of greater susceptibility to decay, disintegration, etc. Similarly, retention by younger rats may be poorer because the immense subsequent brain growth in characteristics such as number of dendritic connections and increasing myelinization may completely mask the original memory, rendering it quite irretrievable. Alternatively, the memory actually may be retrievable given the appropriate retrieval cues, but is not evidenced because of greater susceptibility by younger rats to other relevant factors such as interference or generalization decrement (for a more complete consideration of these and related issues, see Campbell & Spear, 1971).

3. Interference and Forgetting

The influence of PI on retention usually has been conceptualized as limiting retrieval of the more recent learning. But it is plausible that learning preceded by conflicting learning is inherently less stable than other learning—an "eroded" memory trace (cf. Adams, 1967)—and so is lost from storage sooner. In our view, of course, the existing evidence does not favor such an interpretation and the influence of PI is more usefully interpreted as retrieval failure.

D. SPECIFIC MEMORY ATTRIBUTES

This chapter follows others (Bower, 1967b; Underwood, 1969) in assuming that the property of behavior conventionally referred to as a memory actually consists of a number of separate and independent memory attributes. Our experience has indicated that the present position is rarely communicated fully without some explicit comment concerning evidence and implications of this assumption, a task to which we now turn.

By this time it must be understood that details of the attributes of animal memory—their identification, their properties, and the precise means by which a given attribute might become less effective (or more effective) for retrieval of a target attribute—remain matters of largely future concern. In contrast to the evidence for specific functions of comparable attributes in human memory which has been convincingly mobilized by Underwood (1969) in support of this position, applicable animal-memory data appear sparse indeed. Nevertheless this final section will discuss briefly a somewhat obvious implication of the above assumption followed by brief reference to data implicating specific attributes of animal memory.

1. Change in the Retrieval and Discriminative Capacity of Memory Attributes

Gleitman (Chapter 1) has summarized some evidence that certain characteristics of a learning task may be forgotten at different rates. Such evidence implies that the retrieval value of a given memory attribute may decrease independently of the value of associated attributes. A particular attribute may become less discriminable from similar attributes which define other memories, or this particular attribute may no longer be employed for any memory because, e.g., the animal may have matured and switched into a different information processing system; concurrently, another attribute may be entirely unchanged.

If some attributes of a memory are indeed independent, it must be inappropriate to conceptualize behavior of animals as representing "complete forgetting" or "perfect retention." When results appear to indicate complete forgetting or perfect retention, we are almost certainly viewing a consequence of incomplete measurement. At issue here is a matter similar to that concerning the learning-performance distinction, but with important additional features.

Particularly with animals, it is difficult to imagine that any of our current measurement techniques assess anything but partial retention, because most of the memory attributes which might be available to the animal at the retention test simply cannot now be measured. Verbal behavior permits more flexibility in this regard with human memory. Among equally irretrievable verbal units, humans may retrieve more attributes associated with certain items—those on the "tip of the tongue"—than others (Brown & McNeill, 1966). Probably a human could, if asked, retrieve specific information regarding the experimental cubicle in which he originally was trained, the approximate time of his original training, the experimenter and perhaps the experimenter's perfume, even though quite unable to recall a single verbal item which he had learned under these circumstances.

The obvious point is that when an animal chooses correctly 100% of the time during a retention test, we cannot be certain that all memory attributes associated with learning have been aroused; nor can we be sure that no memory attributes have been aroused if the animal chooses randomly. We must infer the proportion of associated attributes available at the retention test from the proportion of animals choosing correctly at a given time or the proportion of correct choices by a single animal over many trials. This method alone surely is inadequate for the identification of the fate of specific attributes. Perhaps direct neural measurement may soon permit assessment and identification of appropriate memory attributes which are aroused during the retention test even though the animal behaves overtly as if no prior training had been given at all. But this would still leave us short of knowing whether this neural activity possesses the capacity for altering the behavior in question. Thus behavioral techniques such as that applied by Steinman (1967) in her "second" retention test (see Gleitman, Chapter 1) may be both immediately and ultimately most useful for an "elemental" analysis of animal memory in terms of specific attributes.

2. Some Attributes of Animal Memory

General principles governing the identification and function of memory attributes used by animals are, of course, not available. Inevitably they will depend upon an interaction between genetic predisposition and early experience along with contemporary factors such as familiarity with the learning environment (e.g., Carlton & Vogel, 1967; Miller, 1969). Many effective memory attributes appear to be "intuitively" obvious to those engaged in animal research, particularly to those investigators well acquainted with the specific species involved. For example, Asratian's "switching" experiments (1965) indicate that the specific experimenter, experimental chamber, time of day at which training or testing is given, and reinforcer magnitude constitute effective memory attributes for dogs. Generally, identification of effective memory attributes may follow the same principles involved in consideration of "stimulus relevance" (see Revusky, Chapter 4). Once identified, the questions remain whether a given attributes serves primarily a discriminative or retrieval function and whether this distinction will indeed be useful in the case of lower animals. I believe such a distinction will be useful at some future time, but current data neither demand nor permit it.

Probably any discriminative stimulus may be shown to function as an effective memory attribute. In addition, there is evidence for the existence of other attributes which do not necessarily exert stimulus control within the learning task but influence memory retrieval nevertheless. Four brief examples of these are presented next.

a. Characteristics of the Apparatus. Gleitman (Chapter 1) has discussed work by Perkins and Weyant, Steinman, the McAllisters and their associates, and Thomas and his associates which indicates that incidental features of the apparatus in which an animal is trained may serve as memory attributes. Specifically, the results of these studies indicate that memory attributes representing certain apparatus features may lose their effectiveness (i.e., may be forgotten) at rates which differ from those of attributes representing the instrumental response. Moreover, data reported by Thomas and Burr (1969) suggest that these sorts of attributes may be reactivated after an otherwise detrimental retention interval by a brief period of exposure to the apparatus. In this experiment, generalization gradients derived from pigeons 24 hours after discrimination training indicated that forgetting had occurred over this period; but this forgetting was eliminated when the retention test was preceded by a 3-minute session in the apparatus accompanied by presentation of the positive stimulus along with the previous reinforcer. These results suggest that the impaired retention of discrimination learning usually found in this situation after a 24-hour interval is caused by the absence of memory attributes representing features of the apparatus and the reinforcer; but a 3-minute exposure to the apparatus and the reinforcer is sufficient to arouse these attributes and thus retrieve the target memory of discrimination learning.

b. Characteristics of the Reinforcer. Spear (1967) has described some instances in which the magnitude of the reinforcer may function as a memory attribute. Briefly, after the reinforcer is changed and behavior yields no effects of the previous reinforcer, reinstatement of appropriate retrieval cues—either by increasing the retention interval or by presenting the original reinforcer—is often sufficient to arouse the memory attribute of the original reinforcer as inferred from the rat's behavior.

c. Emotional Responses. Feigley and Spear (1970) trained weanling and adult rats on a one-way active-avoidance task. The criterion of learning was attained equally rapidly by both age groups. After 1 or 28 days the rats were returned and given an "active-avoidance trial" which differed from the usual in that no shock was delivered if the rat did not avoid within 5 seconds (the rat was permitted as long as 60 seconds to "avoid" before removal from the apparatus). In contrast to the poorer long-term retention by weanlings in terms of trials-to-relearn, the adults had greater first-trial response latencies after 28 days than after 1 day, but the weanlings did not. However, response latencies for weanlings tested after the 28-day interval did not differ from those of experimentally naive rats, while the corresponding latencies of adults were *longer* than those of naive rats. This result was explained by assuming that the adults had retained a general CER better than the specific instrument response, but the weanlings showed relative failure in retention of both components. If this explanation is correct, it implies that memory attributes responsible for the CER and the instrumental response may lose their effectiveness at different rates over a long retention interval.

d. Chemical and Neural Status of the Animal. The above discussion of nonmonotonic changes in retention includes sufficient elaboration on memory attributes which might represent the internal physiological state of the organism at the time of learning. However, because many studies have yielded comparable state-dependent-learning effects only by employing unusually large doses of drugs (e.g., Overton, 1964) there may be reluctance to assign much importance to memory attributes which represent internal events. Perhaps, however, something like "stimulus relevance" (see Revusky, Chapter 4) is applicable in this case; perhaps animals respond most readily to those internal events to which they are most accustomed to responding, or which are most consonant with their specific physiological characteristics.

VI. Conclusion

This chapter has treated the behavioral effects of memory as entirely dependent upon retrieval processes. A variety of experiments, though including a relatively narrow segment of species and their behavior, were shown to be consistent with one view of the way in which memory retrieval might function. The question of whether something like memory decay might influence behavior was not investigated, but it was noted that whatever the effect of memory decay, memory retrieval must remain an important determinant of behavior.

REFERENCES

Adams, J. A. *Human Memory.* New York: McGraw-Hill, 1967.

Amsel, A. Frustrative nonreward in partial reinforcement and discrimination learning: Some recent history and a theoretical extension. *Psychological Review,* 1962, **69,** 306–328.

Amsel, A. Partial reinforcement effects on vigor and persistence. In K. W. Spence & J. T. Spence (Eds.), *The psychology of learning and motivation.* Vol. 1. New York: Academic Press, 1967. Pp. 1–65.

Asratian, E. A. *Compensatory adaptations, reflex activity, and the brain.* Oxford: Pergamon Press, 1965.

Avis, H., & Carlton, P. L. Retrograde amnesia produced by hippocampal spreading depression. *Science,* 1968, **161,** 73–75.

Baum, M. "Reversal Learning" of avoidance response as a function of prior fear conditioning and fear extinction. *Canadian Journal of Psychology,* 1965, **19,** 85–93.

Baum, M. Reversal learning of avoidance response and the Kamin Effect. *Journal of Comparative and Physiological Psychology,* 1968, **66,** 495–497.

Bintz, J. Time-dependent memory deficits of aversively motivated behavior. *Learning and Motivation,* 1970, **1,** 382–390.

Bintz, J., Braud, W. G., & Brown, J. S. Analysis of the role of fear in the Kamin Effect. *Learning and Motivation,* 1970, **1,** 170–176.

Bilodeau, E. A. Retention. In E. A. Bilodeau (Ed.), *Acquisition of skill.* New York: Academic Press, 1966. Pp. 315–350.

Bower, G. H. Verbal Learning. In H. Helson & W. Bevan (Eds.), *Contemporary approaches to psychology.* Princeton, N.J.: Van Nostrand, 1967. Pp. 182–222.(a)

Bower, G. H. A multicomponent theory of the memory trace. In K. W. Spence & J. T. Spence (Eds.), *The psychology of learning and motivation.* Vol. 1. New York: Academic Press, 1967. Pp. 229–325.(b)

Bresnahan, E. L., & Riccio, D. C. Effect of passive avoidance learning upon subsequent active avoidance responding. Paper presented at meetings at the Midwestern Psychological Association, Cincinnati, May, 1970.

Brown, R., & McNeill, D. "The Tip of the Tongue" phenomenon. *Journal of Verbal Learning and Verbal Behavior,* 1966, **5**, 325–337.

Brush, F. R., & Levine, S. Adrenocortical activity and avoidance behavior as a function of time after fear conditioning. *Physiology & Behavior,* 1966, **1**, 309–311.

Brush, F. R., Myer, J. S., & Palmer, M. E. Effects of kind of prior training and intersession interval upon subsequent avoidance learning. *Journal of Comparative and Physiological Psychology,* 1963, **56**, 539–545.

Campbell, B. A. Developmental studies of learning and motivation in infraprimate mammals. In H. W. Stevenson, E. H. Hess, & R. L. Rheingold (Eds.), *Early behavior: Comparative and developmental approaches.* New York: Wiley, 1967. Pp. 43–71.

Campbell, B. A., & Spear, N. E. Ontogeny of memory. *Psychological Review,* 1971, manuscript submitted.

Capaldi, E. J. Partial reinforcement: A hypothesis of sequential effects. *Psychological Review,* 1966, **6**, 321–322.

Capaldi, E. J. A sequential hypothesis of instrumental learning. In K. W. Spence & J. P. Spence (Eds.), *The psychology of learning and motivation.* Vol. 1. New York: Academic Press, 1967. Pp. 67–156.

Carlton, P. L. Brain-acetylcholine and inhibition. In J. T. Tapp (Ed.), *Reinforcement and behavior.* New York: Academic Press, 1969. Pp. 286–327.

Carlton, P. L., & Vogel, J. R. Habituation and conditioning. *Journal of Comparative and Physiological Psychology,* 1967, **63**, 348–351.

Ceraso, J. The interference theory of forgetting. *Scientific American,* 1967, **217**, 117–124.

Chiszar, D. A., & Spear, N. E. Proactive interference in retention of nondiscriminative learning. *Psychonomic Science,* 1968, **12**, 87–88.(a)

Chiszar, D. A., & Spear, N. E. Proactive interference in a T-maze brightness-discrimination task. *Psychonomic Science,* 1968, **11**, 107–108.(b)

Chiszar, D. A., & Spear, N. E. Stimulus change, reversal learning and retention in the rat. *Journal of Comparative and Physiological Psychology,* 1969, **69**, 190–195.

Cole, M., & Hopkins, D. Proactive interference for maze habits in the rat. *Psychonomic Science,* 1968, **10**, 365–366.

Crowder, R. G., Cole, M., & Boucher, R. Extinction and response competition in original and interpolated learning of a visual discrimination. *Journal of Experimental Psychology,* 1968, **77**, 422–428.

D'Amato, M. R., & Jagoda, H. Effects of extinction trials on discrimination reversal. *Journal of Experimental Psychology,* 1960, **59**, 254–260.

Denny, M. R., & Ditchman, M. E. The locus of maximal "Kamin Effect" in rats. *Journal of Comparative and Physiological Psychology,* 1962, **55**, 1069–1070.

Desiderato, O., Butler, D., & Meyer, C. Changes in fear generalization gradients as a function of delayed testing. *Journal of Experimental Psychology,* 1966, **72**, 678–682.

Donahoe, H. W., McCroskery, J. H., & Richardson, W. K. Effects of context on the post discrimination gradient of stimulus generalization. *Journal of Experimental Psychology,* 1970, **84**, 58–63.

Ehrenfreund, D., & Allen, J. D. Perfect retention of an instrumental response. *Psychonomic Science,* 1964, **1**, 347–348.

Estes, W. K. Statistical theory of spontaneous recovery and regression. *Psychological Review,* 1955, **62**, 154.

Estes, W. K. The statistical approach to learning theory. In S. Koch (Ed.), *Psychology: A study of a science.* Vol. 2. New York: McGraw-Hill, 1959. Pp. 380–491.

Feigley, D. A., & Spear, N. E. Effect of age and punishment condition on long-term retention by the rat of active- and passive-avoidance learning. *Journal of Comparative and Physiological Psychology,* 1970, **73,** 515–526.

Gleitman, H., & Jung, L. Retention in rats: The effect of proactive interference. *Science,* 1963, **142,** 1683–1684.

Gonzalez, R. C., Behrend, E. R., & Bitterman, M. E. Reversal learning and forgetting in bird and fish. *Science,* 1967, **158,** 519–521.

Haltmeyer, G. C., Denenberg, V. H., & Zarrow, M. X. Modification of the plasma corticosterone response as a function of infantile stimulation and electric shock parameters. *Physiology & Behavior,* 1967, **2,** 61–63.

Hili, W. F., Cotton, J. W., & Clayton, K. N. Effect of rewarded and nonrewarded incorrect trials on T-maze learning. *Journal of Comparative and Physiological Psychology,* 1963, **56,** 489–496.

Honig, W. K. Attentional factors governing the slope of the generalization gradient. In R. M. Gilbert & N. S. Sutherland (Eds.), *Animal discrimination learning.* New York: Academic Press, 1969. Pp. 35–62.

Hughes, R. A. Retrograde amnesia produced by hippocampal injection of potassiufn chloride: Gradient of effect and recovery. *Journal of Comparative and Physiological Psychology,* 1969, **68,** 637–644.

Kamin, L. J. The retention of an incompletely learned avoidance response. *Journal of Comparative and Physiological Psychology,* 1957, **50,** 457–460.

Kamin, L. J. Retention of an incompletely learned avoidance response: Some further analysis. *Journal of Comparative and Physiological Psychology,* 1963, **56,** 713–718.

Kamin, L. J., Brimer, C. J., & Black, A. H. Conditioned suppression as a monitor of fear of the CS in the course of avoidance training. *Journal of Comparative and Physiological Psychology,* 1963, **56,** 497–501.

Kehoe, J. Effects of prior and interpolated learning on retention in pigeons. *Journal of Experimental Psychology,* 1963, **65,** 537–545.

Keppel, G. Retroactive and proactive inhibition. In T. R. Dixon & D. L. Horton (Eds.), *Verbal behavior and general behavior theory.* Englewood Cliffs, N.J.: Prentice Hall, 1968. Pp. 172–213.

King, M. G. Stimulus generalization of conditioned fear in rats over time: Olfactory cues and adrenal activity. *Journal of Comparative and Physiological Psychology,* 1969, **69,** 590–600.

Klein, S. B., & Spear, N. E. Influence of age on short-term retention of active-avoidance learning in rats. *Journal of Comparative and Physiological Psychology,* 1969, **69,** 583–589.

Klein, S. B., & Spear, N. E. Forgetting by the rat after intermediate intervals ("Kamin Effect") as retrieval failure. *Journal of Comparative and Physiological Psychology,* 1970, **71,** 165–170.(a)

Klein, S. B., & Spear, N. E. Reactivation of avoidance-learning memory in the rat after intermediate retention intervals. *Journal of Comparative and Physiological Psychology,* 1970, **72,** 498–504.(b)

Koppenaal, R. J., & Jagoda, E. Proactive inhibition of a maze position habit. *Journal of Experimental Psychology,* 1968, **76,** 664–688.

Levine, S., & Brush, F. R. Adrenocortical activity and avoidance learning as a function of time after avoidance training. *Physiology & Behavior,* 1967, **2,** 385–388.

Levine, S., & Mullins, F. F., Jr. Hormones in infancy. In G. Newton & S. Levine (Eds.), *Early experience and behavior.* Springfield, Ill.: Thomas, 1968. Pp. 168–197.

Lewis, D. J. Sources of experimental amnesia. *Psychological Review,* 1969, **76,** 461–472.

Lewis, D. J., & Maher, B. A. Neural consolidation and electro-convulsive shock. *Psychological Review,* 1965, **72,** 225–239.

McAllister, D. E., & McAllister, W. R. Incubation of fear: An examination of concept. *Journal of Experimental Research in Personality,* 1967, **3**, 80–90.

McAllister, D. E., & McAllister, W. R. Forgetting of acquired fear. *Journal of Comparative and Physiological Psychology,* 1968, **65**, 352–355.

McGaugh, J. L. Time-dependent processes in memory storage. *Science,* 1966, **153**, 1351–1358.

Mackintosh, N. J., McGonigle, B., Holgate, V., & Vanderver, F. Factors underlying improvement in serial reversal learning. *Canadian Journal of Psychology,* 1968, **22**, 85–95.

Maier, S. F., Allaway, T. A., & Gleitman, H. Proactive inhibition in rats after prior partial reversal: A critique of the spontaneous recovery hypothesis. *Psychonomic Science,* 1967, **9**, 63–65.

Maier, S. F., & Gleitman, H. Proactive interference in rats. *Psychonomic Science,* 1967, **7**, 25–26.

Mellgren, R. L. Demonstrations of positive contrast in the runway. Paper presented at meetings of the Midwestern Psychological Association, Cincinnati, May, 1970.

Miller, R. R. Effects of environmental complexity on amnesia induced by electro-convulsive shock. Unpublished doctoral dissertation, Rutgers University, 1969.

Overton, D. State-dependent or "dissociated" learning produced with pentobarbital. *Journal of Comparative and Physiological Psychology,* 1964, **57**, 3–12.

Postman, L. The present status of interference theory. In C. N. Cofer (Ed.), *Verbal learning and verbal behavior.* New York: McGraw-Hill, 1961. Pp. 152–196.

Postman, L., Stark, K., & Fraser, J. Temporal changes in interference. *Journal of Verbal Learning and Verbal Behavior,* 1968, **7**, 672–694.

Razran, G. H. S. Extinction, spontaneous recovery and forgetting. *American Journal of Psychology,* 1939, **52**, 100–102.

Seligman, M. E. P., Maier, S. F., & Solomon, R. L. Unpredictable and uncontrollable aversive events. In F. R. Brush (Ed.), *Aversive conditioning and learning.* New York: Academic Press, 1971, in press.

Selye, H. The general adaptation syndrome and the diseases of adaptation. *Endocrinology,* 1946, **36**, 2, 117–230.

Skinner, B. F. Are theories of learning necessary? *Psychological Bulletin,* 1950, **57**, 193–216.

Spear, N. E. Retention of reinforcer magnitude. *Psychological Review,* 1967, **64**, 216–234.

Spear, N. E. Lapses and losses in retention by the rat. Paper presented at Symposium on Animal Retention, meetings of the Midwestern Psychological Association, Chicago, May, 1969.

Spear, N. E. Verbal learning and retention. In M. R. D'Amato. *Experimental psychology: Methodology, psychophysics and learning.* New York: McGraw-Hill, 1970. Pp. 543–638.

Spear, N. E., Hill, W. F., & O'Sullivan, D. J. Acquisition and extinction after initial trials without reward. *Journal of Experimental Psychology,* 1965, **69**, 25–29.

Spear, N. E., Klein, S. B., & Riley, E. P. The Kamin Effect as "state-dependent": Memory-retrieval failure in the rat. *Journal of Comparative and Physiological Psychology,* 1971, **74**, 416–425.

Spear, N. E., & Spitzner, J. H. Effects of initial nonrewarded trials: Factors responsible for increased resistance to extinction. *Journal of Experimental Psychology,* 1967, **74**, 525–537.

Spear, N. E., & Spitzner, J. H. Residual effects of reinforcer magnitude. *Journal of Experimental Psychology,* 1968, **77**, 135–149.

Spevack, A. A., & Suboski, M. D. Retrograde effects of electroconvulsive shock on learned responses. *Psychological Bulletin,* 1969, **72,** 66–76.

Spitzner, J. H., & Spear, N. E. Studies of proactive interference in retention of the rat. Paper presented at the meetings of the Psychonomic Society, St. Louis, October, 1967.

Steinman, F. Retention of alley brightness in the rat. *Journal of Comparative and Physiological Psychology,* 1967, **64,** 105–109.

Suboski, M. D., Marquis, H. A., Black, M., & Platenius, P. Adrenal and amygdala function in the incubation of aversively conditioned responses. *Physiology & Behavior,* 1970, **5,** 283–289.

Thomas, D. R., & Burr, D. E. S. Stimulus generalization as a function of the delay between training and testing procedures: A reevaluation. *Journal of the Experimental Analysis of Behavior,* 1969, **12,** 105–109.

Underwood, B. J. Degree of learning and the measurement of forgetting. *Journal of Verbal Learning and Verbal Behavior,* 1964, **3,** 112–129.

Underwood, B. J. Attributes of memory. *Psychological Review.* 1969, **76,** 559–573.

Underwood, B. J., & Freund, J. S. Effect of temporal separation of two tasks on proactive inhibition. *Journal of Experimental Psychology,* 1968, **78,** 50–54.

Weiss, J. M., McEwen, B. S., Silva, M. T. A., & Kalkut, M. F. Pituitary-adrenal influences on fear responding. *Science,* 1969, **163,** 197–199.

Chapter 3

MEMORY AND LEARNING:
A SEQUENTIAL VIEWPOINT[1]

E. J. Capaldi

The sequential hypothesis, in common with several other approaches to instrumental learning, places considerable emphasis on internal organismic stimuli which result from the occurrence of reward and of nonreward. How these stimuli are produced is a matter of considerable contemporary concern. In practice the nature or characteristics of the "stimulus-producing mechanism" are inferred as follows: The first step is to isolate experimental variables which are thought to be responsible for the appearance of the internal stimuli. The next step is to describe a mechanism thought to be implied by these variables, and finally one seeks to confirm these propositions experimentally. The sequential hypothesis suggests that the mechanism can profitably be conceptualized as memory. What is meant by this may be illustrated by a simple example. On a particular trial an animal is nonrewarded. Since the sequential hypothesis endows the animal with an impressive memory capacity, it is assumed that on the immediately subsequent trial the animal will remember being nonrewarded. That is, on the subsequent trial the animal's internal stimulus state will be that having the characteristics of nonreward rather than, let us say, reward. If on a given trial the animal remembers nonreward and is on that trial rewarded, then stimuli characteristic of nonreward will become conditioned to, and thus acquire control over, the instrumental reaction (e.g., Capaldi, 1966).

The sort of experimental data which would lead one to this conception are rather straightforward, and while much of it has accumulated only recently, some of it has been available for years. Consider, for example, a finding reported by Grosslight and his collaborators in a series of experiments using human and animal subjects (see Grosslight, Hall, & Murnin, 1953). Grosslight and his collaborators typically employed different sequences of rewarded (R) and nonrewarded (N) trials. One group might receive each day the three-trial reward

[1] This research was supported in part by National Institute of Child and Health Development Grant HD 04379.

schedule RNR, the other group the schedule RRN. Note that these groups are equated for such variables as number of rewarded trials, number of nonrewarded trials, total number of trials, percentage of reward, and so on. And while in some of the experiments reported by Grosslight and collaborators, the animal might form a discrimination between rewarded and nonrewarded trials (because of the predictable or regular trial sequence), in other experiments there was no evidence that such a discrimination had been formed (see Grosslight & Radlow, 1957). In any case, the presence or absence of a discrimination of pattern did not appear to affect the major finding obtained in this series of investigations, which was that partial reinforcement increased resistance to extinction following RNR training, but not following RRN training. Other investigations reported by a variety of individuals (see Capaldi, 1967) have seemed to establish that the partial reinforcement effect will occur only if nonrewarded trials are followed by rewarded trials (N–R transitions).

The sequential interpretation of this general finding in groups trained RNR is relatively straightforward. On the third or last trial of the day the animal remembers being nonrewarded on the previous trial. Since the third trial is a rewarded one, the stimulus characteristic of nonreward is conditioned to the instrumental reaction. Extinction, of course, consists of a series of nonrewarded trials. And resistance to extinction may be considered as an index of the capacity of nonreward-related stimuli to evoke the instrumental reaction. Quite evidently, the capacity of the stimuli related to nonreward to evoke the response in groups trained on an RNR schedule is considerable, since groups trained in this manner show very great resistance to extinction. Groups trained on an RRN schedule, according to this view, do not show increased resistance to extinction because nonreward stimuli have not occurred on subsequent rewarded trials, and thus have acquired little capacity to evoke the instrumental reaction. The more complete explanation is more complicated, as we shall see later, but for present purposes the above account is adequate.

It might be asked whether it must be assumed that animals trained on an RRN schedule have no awareness whatever of being nonrewarded. Surely animals are capable of learning that nonreward sometimes occurs in the apparatus. According to this view, the animal would surely learn, generally speaking, that reward is not always forthcoming in the apparatus and so, on given trials, would be in some kind of nonreward stimulus state, even if the previous trial was not nonrewarded. Consider, for example, Trial 2 of the RRN schedule. According to the memory assumption, on Trial 2 (which, of course, follows a rewarded trial) the animal will be in a stimulus state characteristic of reward. But if, as suggested above, the animal learns that nonreward is not always forthcoming in the apparatus, it would be in some kind of a nonreward stimulus state as well. Granting this, as we do, it might be asked why are these nonreward stimuli not conditioned to the reaction, and why then do they not support increased resistance to extinction? We would reply as follows. There seems to be no reason why such stimuli would not be conditioned to the instrumental reaction. However, it seems to us undeniable that the conditioning of such stimuli to the instrumental reaction fails to increase the animal's capacity to withstand extinction in view of findings such as those reported by Grosslight and his collaborators, and others. The reason that such stimuli fail to increase resistance to extinction is perhaps straightforward. In extinction all nonrewarded trials except the first are preceded by nonrewarded trials. Thus on every extinction trial save the first the animal remembers nonreward. Let us assume first that the nonreward stimuli produced by the animal's more or less general expectation of being nonrewarded fail to supply the stimuli produced by the animal's memory of being nonrewarded with considerable generalized associative strength. Let us assume further that extinction is an index mainly of the associative capacity of remembered nonreward stimuli to evoke the instrumental reaction. It then becomes understandable that an animal trained on an RRN schedule would not show increased resistance to extinction. Perhaps there are other, better explanations than the one offered above. Whatever the explanation, however, it must take into account the fact that some sorts of instrumental behaviors, of which extinction is one of the better known examples, appear to be primarily regulated by remembered nonreward, and that other classes of stimuli related to nonreward appear to exercise little or no direct influence over these behaviors.

The sequential hypothesis grew out of what is called the Hull-Sheffield hypothesis (Hull, 1952; Sheffield, 1949). This hypothesis conceived of internal reward- and nonreward-related stimuli as traces or aftereffects which persisted, as it were, from one trial to the next. The internal stimuli could be described, then, as sensory-peripheral feedback which persisted for a time and then faded and disappeared. According to this view, the animal would be in a nonreward stimulus state only following a nonrewarded trial. Thus, the sequence of rewarded and nonrewarded trials was for the Hull-Sheffield hypothesis an important matter. The Hull-Sheffield hypothesis, when first proposed, was widely regarded as *the* solution to certain classes of recalcitrant problem, particularly those associated with the effects of partial reward, which did not seem to be understandable in other ways. However, despite the acclaim, the Hull-Sheffield hypothesis was, for a variety of reasons which do not concern us at the moment, eventually rejected and abandoned. It is not surprising, perhaps, that many of the historical objections to the Hull-Sheffield hypothesis have been directed to the sequential approach. Accordingly, it might be thought that a proposition central to both the sequential view and the Hull-Sheffield hypothesis, that the specific trial sequence is important to the nature of the stimulus control established over responding, remains controversial. This is no longer the case. The importance of trial sequence in the overall stimulus control process is acknowledged by many (e.g., Amsel, 1967; Gonzalez & Bitterman, 1969; Logan, 1968; Spence, Platt, & Matsumoto, 1965). However, the tendency is to see sequence as important in the stimulus control process only when trials are massed and not when trials are spaced. That is, many remain unconvinced that memory, rather than a sensory-peripheral feedback mechanism, provides the stimuli which ultimately exercise control over performance.

The sequential approach shares three important characteristics with other contemporary approaches to instrumental learning: all attempt to specify both the characteristics of reward- and nonreward-related stimuli and the mechanism which produces these stimuli, and they try to elucidate the nature of the conditioning process whereby these stimuli become associated with the appropriate instrumental responses. While these matters are closely interrelated, they can for purposes of analysis be separated. This chapter will deal primarily with the second of the questions outlined above, the nature of the mechanism producing the internal reward- and nonreward-related stimuli which ultimately become conditioned to, and acquire control over, the instrumental reaction. The position suggested here, of course, is that the internal stimuli can profitably be conceptualized as memories or, to put it slightly differently, that the internal stimuli are produced by a mechanism which follows the laws of memory.

In addition to sensory-peripheral feedback and memory, one other mechanism which has been alluded to above has been prominently identified as producing the internal stimuli which ultimately come to regulate instrumental responding. The stimuli are produced, according to this view, by reactions classically conditioned to apparatus cues (Amsel, 1958; Spence, 1960). Earlier, when we spoke of the animal's learning generally that reward was not always forthcoming in the apparatus, it was the classical conditioning mechanism that we had in mind. Regarding these alternatives to memory, the following may be said: Research from our laboratory as yet unpublished appears to suggest rather clearly that short-term stimuli which could be described as fading traces or aftereffects (the mechanism producing these stimuli is not clear to us) do, under certain conditions, exercise a significant control over instrumental responding. Moreover, it is rather difficult to imagine that classical conditioning or some similar process is not involved in the production of stimuli which come ultimately to exercise some sort of control over instrumental responding. It seems probable to us, however, that the role (or roles) which classical conditioning plays in instrumental learning has not as yet been precisely isolated and identified. Certainly this role is currently undergoing reevaluation. Thus, massed-trial extinction, to select what may be a less controversial example, while once considered to be under the control of stimuli provided by the classically conditioned reaction, is now regarded by many as being regulated by sequential variables (Amsel, 1967; Gonzalez & Bitterman, 1969; Logan, 1968). In our view, then, those aspects of instrumental learning which are primarily regulated by sensory-peripheral feedback from classically conditioned reactions remain to be isolated and identified.

Little will be said here concerning the properties or characteristics of reward- and nonreward-related stimuli. The identification and precise description of the characteristics of the stimuli regulating instrumental behavior are quite clearly two of the more important tasks facing the learning psychologist. The truth of this proposition could not be more amply illustrated than by noting the increase in our understanding which has been generated by Amsel's (1958) identification of nonreward as being under some circumstances frustrating. Much instrumental behavior does not seem understandable in the absence of the frustration assumption. This, of course, is widely recognized. The preoccupation of this chapter, however, is with the nature of the mechanism producing the reward- and nonreward-related stimuli, and not with the characteristics of the stimuli so produced. While the two questions are closely related, they are nevertheless separable.

The plan of this chapter is as follows: In Part I, an attempt will be made to describe why the need for an internal stimulus approach arose within learning theory, what sorts of internal stimulus mechanisms have been proposed, how the hypothesis of a memory mechanism deals with various objections which have been raised against it, to what circumstances it traces these objections, what the advantages of a memory approach seem to be, and several other related matters. Part II will provide a description of some recent research which appears to dispose effectively of some classical objections to a memory approach and which clarifies the operation of the memory mechanism, particularly when external stimulus conditions vary as in the discrimination learning situation. Some of the research described in Part II is quite unfavorable to other current internal stimulus approaches.

PART I

I. Historical Background

Hull attempted to explain the phenomena of learning in his monumental book *The Principles of Behavior* (1943) by postulating a learned connection, a habit, between instrumental responses and external stimuli. While Hull was certainly not oblivious to internal stimuli of various sorts, it is clear that the burden of theoretical analysis in the *Principles* fell upon external stimuli and their relation to the instrumental reaction. Thus, to take a straightforward example, Hull reasoned that the larger the reward magnitude, the stronger the external stimulus–instrumental response connection or habit, and, accordingly, the faster the running. For a variety of excellent reasons, this sort of analysis simply will not do. For example, partial reinforcement produces greater resistance to extinction than consistent reinforcement, yet under partial reward there are fewer conditionings of external stimuli to the instrumental reaction than under consistent reward. Learning theorists slowly came to see what was required to deal with certain newer forms of phenomena which were then coming under investigation. Many concluded that in order to deal with such data, stimuli had to be put into the organism, either into his mouth or legs (peripheral) or into his head or nervous system (central). Let us describe briefly and specifically how and why this change came about.

Prior to 1940, Hull and learning theorists generally dealt with what may be called constant conditions of reward. In such cases, if the animal received, for example, a 10-pellet reward on the initial trial, it received a 10-pellet reward on all subsequent trials. An external stimulus model works fairly well for conditions of constant reward—indeed, many dramatically different sorts of models will work about equally well in this situation. However, if one varies the conditions of reward from trial to trial, an external stimulus approach breaks down so unmistakably that few would care to defend it. It was at this point that internal stimuli were brought into the picture. Hull, with his mind and heart apparently still in the external stimulus approach, moved only part of the way toward recognizing the importance of internal stimuli in 1952. Spence moved slightly further in this direction in 1956, but hardly far enough to be effective; the 1956 system is not useful for dealing with varied reward, perhaps in part because it did not take varied reward seriously enough.

The decision to theorize in terms of internal stimuli immediately raises a variety of new and perplexing problems, so it should not be surprising that progress along this line has been, and continues to be, slow and tentative. One needs, for example, to identify the mechanism which provides or produces the internal stimuli. It also must be determined what sorts of internal stimuli are provided. Moreover, an identification of the experimental conditions responsible for the appearance of the internal stimuli must be attempted. Clearly the nature of this functional relationship, as well as that between the internal stimulus and the response, must be described. Furthermore, it must be determined what the properties of the internal stimuli may be. Still other sorts of questions must be decided. In approximately the last 10 years at least some of the questions raised in this section have been the object of considerable research and controversy. In this context, the slow pace at which our ability to deal with varied reward has developed is hardly surprising—it could not have been otherwise. In our opinion, constant reward data could hardly have suggested, let alone provided or prepared us for, the kinds of principles required to explain learning phenomena. The error of learning theory in the 1930's and 1940's, as we see it, was in collecting the wrong kinds of data, a tendency which, from the viewpoint of sequential theory, persists to some extent even now.

II. Varieties of Models

Not surprisingly, some of the issues described above are decided, at least to a degree, once one settles upon a mechanism for providing the internal stimuli. Sheffield (1949), for example, postulated a rather simple sensory-peripheral or stimulus trace mechanism which went far toward deciding either directly or by implication at least some of the issues raised in the preceding section. One of Sheffield's suggestions was that following a rewarded trial the animal might have left-over food in its mouth, a condition largely absent following a nonrewarded trial. Following the two sorts of trial, then, the animal's internal stimulus state would tend to differ. Let us now direct our attention to two other radically different sorts of internal stimulus mechanisms which have been proposed.

One is a learning mechanism. It was suggested originally by Hull in 1931 as the anticipatory response conception (r_g–s_g). In more recent years it has been updated and extended by Amsel (1958) and later by Spence (1960) into what has been called the frustration hypothesis. Despite its antiquity, the r_g model as elaborated in the frustration hypothesis proved to be capable of dealing quite effectively with some aspects of the new varied reward phenomena. What interests us about this model is not its emphasis upon frustration, but rather the mechanism it employs for providing the animal with internal stimuli, which is a learning mechanism. On making the assumption that the animal is provided with internal stimuli through a learning process, the r_g model is committed, as it were, to the implications of this approach in every learning situation. However, even its most staunch adherents would not care to defend that proposition for all conceivable cases—nor, as we shall see, have they done so. In varied reward situations internal stimuli do not seem to be produced by something so cumbersome, conservative, or slow to react and adjust as a learning mechanism. We shall emphasize this point again and again; indeed, this point is the heart of the chapter.

Let us see how the frustration model works in general. As a function of goal box reward, earlier portions of the apparatus (S_A) come to evoke an anticipatory reaction and an associated response-produced stimulus (S_A–r_g–s_g). The r_g reaction as normally conceived is acquired slowly. The internal stimulus produced by r_g, or s_g, acquires the capacity to evoke the approach reaction (s_g–R_A). Once r_g is sufficiently established, goal box nonreward elicits a frustrative reaction (RF). Over trials, frustration also becomes anticipatory (S_A–r_f–s_f) and s_f like s_g acquires the capacity to evoke approach (s_f–R_A). The s_f–R_A connection is thought to be the mechanism of the partial reinforcement effect (PRE); it is responsible for the PRE since the animal learns to approach in the presence of frustration (s_f–R_A). A consistently rewarded animal will not learn to approach in the presence of frustration, and so will extinguish more rapidly than the partially rewarded animal. Clearly, the animal's internal stimulus environment lags considerably behind the prevailing conditions of reward in the anticipatory response model. Thus as the frustration hypothesis was originally conceived (as we shall see, it has since been rather radically modified) the internal stimulus s_g does not appear until a number of rewarded trials have been experienced, and the internal stimulus s_f does not appear until even later in the reward–nonreward sequence.

On this basis Amsel (1958) took as evidence for the frustration formulation the failure of a PRE to appear following 24 acquisition trials—it was thought that the mechanism of the PRE, s_f–R_A, had not yet been established. In other places it has been suggested that the mechanism of the PRE, s_f–R_A, was not yet established following as many as 60 training trials (Amsel & Ward, 1965). One more example, perhaps, will serve to provide a flavor of the degree of commitment to "gradual internal stimulus change" traditionally favored by the anticipatory response formulation. Thanks largely to the work of McCain (1966), it is by now no longer disputed that a PRE will occur following a very small number of trials, perhaps as few as two, i.e., following the sequence nonreward–reward. Black and Spence (1965), who had in mind about a 10-trial PRE (see Spence *et al.,* 1965), dealt with this information as follows: The early trial PRE was ascribed to the sorts of sequential mechanisms recommended by Capaldi (1964). Later in training, however, s_f–R_A, i.e., frustration, comes into play and the sequential mechanism becomes more or less irrelevant. Thus, in their effort to deal with the small trial PRE, Black and Spence, rather than attempting to modify the gradual internal stimulus change assumption fostered by the anticipatory response approach, preferred to postulate two sorts of PRE, an "early trial" PRE and a "late trial" PRE, each controlled by different processes and variables. Clearly, then, they considered a two-process explanation of instrumental learning phenomena a less radical alternative than that of modifying the gradual internal stimulus change assumption fostered by the r_g approach.

We turn now to a rather different sort of internal stimulus mechanism. Our own preference is to assume that the animal is provided with internal stimuli on the basis of a mechanism best conceptualized as a memory mechanism. Originally our preference for this assumption arose out of our understanding of the implications of various sorts of experimental data. Lately this preference has been reinforced and increased by the realization, which is still growing, that the memory mechanism leads to a kind of learning theory which is unique both in its structure and in its implications. Let us see what it means to say that the animal's internal stimuli are best characterized as memories. Unlike Sheffield, we would hardly maintain that the internal stimuli or traces are at the periphery, or in the mouth, but that they are deeper, in the nervous system. Also, a trace conception implies that once the internal stimuli are initiated, e.g., by food in the mouth, they persist and thus are available whatever the conditions of external stimulation. Thus, such stimuli would be "in the animal" no matter if it is returned to the home cage, if it remains in the intertrial interval box, or if it is put back into the apparatus. Memories, on the other hand, are not independent of external stimuli, but rather depend critically upon external stimulation. Thus a rat shocked in situation X will have occasion to remember such when returned to X, or, at least, to a similar situation. Nor are memories time-dependent; traces fade, but memories return when the situation which initiated them is re-presented. In the interim they are absent, unless provoked by other memories themselves initiated by external stimulation. In contrast to r_g, memories are not learned; classical conditioning is considered to be far too conservative a mechanism for altering the animal's internal stimulus environment. According to the sequential model, the animal will remember a single nonreward on being placed again in the apparatus. Or, if he was rewarded, he will remember this fact on a subsequent trial. Within a learning framework many trials are required for such anticipations to develop, but in the memory model they occur from the outset of training. Within the learning model, variations in the internal stimulus complex are usually small from trial to trial. In the memory model, such variations can occur rapidly, and a single trial is sufficient to produce them. Equally important, the trial-to-trial variations in the stimulus complex can be large or extensive.

III. Memories as Internal Stimuli: Objections to This View

The rapid internal stimulus change proposition, which is at the heart of the present approach, is extremely difficult for many individuals to accept, especially when trials are widely spaced. It should be recognized that cognitive theorists no less than r_g theorists tend to think in terms of slow and gradual strengthening of expectancy over trials. Even theorists in the "one-trial learning" tradition may be reluctant to assume that internal stimuli can change rapidly, although an assumption to this effect could easily be fitted into their framework. Why this reluctance? Primarily we think it represents a legacy from the field's preoccupation in the 1940's with conditions of constant reward. Gradual change in performance in constant reward situations came to mean, understandably enough, gradual changes in r_g or expectancy. This view still persists; in its current form it is most popularly represented in the r_g, anticipatory response approach. While the gradual change position is a possible interpretation, it is by no means a necessary one. Indeed, varied reward situations suggest, quite in contrast, that the gradual change view is neither useful nor feasible.

Another comment is in order here. Even gradual change theories suggest rapid internal stimulus change under some circumstances and, as we shall see, such models are being forced increasingly to adopt this position. To take one example, an animal shifted from large reward to small reward (e.g., Crespi, 1944) is expected on the next trial to be frustrated or whatever—to be in a radically different stimulus state. Within the context of gradual change models, rapid stimulus change of this sort is the exception, of course, and not the rule. To a sequential theorist, cases of this type only point up how strong is the prejudice against the notion of rapid and substantial trial-to-trial variations in the internal stimulus complex. Rapid change appears to be accepted only as a last resort and, even then, on a case-by-case basis.

A. GRADUAL CHANGE NOT A NECESSARY INTERPRETATION

In this example let the animal be consistently rewarded in a straight alley (constant reward). We will attempt to show here that even for cases of constant reward, the gradual change hypothesis is not necessary. Later on, using varied reward, an attempt will be made to show that the gradual change hypothesis does not seem feasible.

In the consistent reinforcement situation according to, say, the r_g hypothesis, the animal must learn two things—to expect food and to perform the instrumental reaction. In technical language, the animal learns $S_A - r_g$ and $s_g - R_A$. Improvement is assumed to be gradual because it takes several trials to establish not only $S_A - r_g$, but also to establish $s_g - R_A$. So much for the learning model.

Let S^R symbolize internalized (remembered) reward.[2] In the sequential model the appearance of S^R is the work of memory, not learning. All that is required in order for S^R to appear is (1) that reward have occurred on the previous trial, and (2) that the animal be returned to the same or a similar external stimulus situation. That S^R is there in full strength, as it were, in no way militates against gradual improvement in performance. In common sense terms, while S^R may be there on the immediately subsequent trial, the animal does not know what to do about it—jump, run, roll over, play dead, groom, or whatever. Gradual improvement is the work of learning or $S^R - R_A$. In the sequential model, then, learning $S^R - R_A$ generates the slowly improving performance curve which is typically observed. From a sequential point of view, gradual change models of various sorts incorporate a fundamental confusion; they confuse the problem of memory with the problem of learning. In a sequential context, memory $(S_A - S^R)$ and learning $(S^R - R_A)$ can and must be separated. It should be indicated in passing that the animals remember things other than reward (S^R) or nonreward (S^N). For example, animals remember what response they made on the previous trial, the time interval between trials, as well as other things (e.g., Capaldi, 1967; Capaldi, Leonard & Ksir, 1968b; Capaldi & Minkoff, 1966). Differences between learning models and memory

---

[2] The reader should distinguish carefully between this meaning and the use of S^R as a symbol for a reinforcing stimulus in the area of operant behavior.

models of internal stimulus change turn out to be rather profound. This can be illustrated when each type of model attempts to characterize what it is that the organism learns. We turn briefly to a consideration of this topic.

IV. What Is Learned: Two Points of View

Consider two points of view involving gradual stimulus change, one a stimulus–response formulation (Amsel, 1958), the other a cognitive formulation (Lawrence & Festinger, 1962). Both assume that the animal is relatively insensitive to the prevailing conditions of reward. According to Amsel, what the animal learns in a varied reward situation is to approach in the face of frustration $(s_f\text{–}R_A)$ and according to Lawrence and Festinger the animal slowly accumulates extra attractions. Given that the animal is insensitive to the prevailing conditions of reward, "what is learned" necessarily depends, according to theorists of this persuasion, on "numbers variables" (e.g., number of rewards or nonrewards or both), and animals which experience the same number of rewards or nonrewards will necessarily learn the same things. That this view cannot be taken literally even in the context of these theories should perhaps be evident. For example, to choose an extreme case, an animal experiencing 20 nonrewards then followed by 20 rewards is bound even in gradual change contexts (as their proponents would admit) to have learned something different from an animal experiencing 20 rewards followed by 20 nonrewards. A discussion of how and to what extent number variables must necessarily be influenced by sequence variables even in the context of views of the frustration and dissonance type would be interesting and informative. However, it lies beyond the scope of this chapter; in any case the main outlines of such a discussion are probably clear. We shall merely indicate, then, that Lawrence and Festinger, for example, suggest that resistance to extinction is an increasing function of number of nonrewards only—the greater the number of nonrewards the greater the accumulated extra attractions, and, therefore, the greater the resistance to extinction. Further, Surridge and Amsel (1966), in considering intertrial reinforcement, have emphasized the importance of number of nonrewards, although they have not systematically integrated the effects of this variable into the overall frustration analysis. According to views of this type, what the animal learns is to be characterized in terms of "strength" or "intensity" of such states as frustration or expectancy, strength and intensity being increasing functions of number of rewards or nonrewards or both.

From the point of view of the sequential hypothesis, the organism's extreme sensitivity to the prevailing conditions of reward has other implications for the question of what is learned. According to this hypothesis, very little that is useful can be inferred about what has been learned if one knows only how many rewards or nonrewards or both have occurred. One must know the sequence of the rewards and nonrewards. Fundamentally, sequence determines what is learned—two animals experiencing different sequences will necessarily learn different things, perhaps radically different things, depending, of course, upon how different the schedules are. Let us attempt to illustrate this point with a simple schedule of large (L) and small (S) rewards.

In an investigation by Leonard (1969) employing two trials a day, one group was trained SL (Trial 1, small reward; Trial 2, large reward), the other group being trained LS. For the SL animals, small reward is remembered on Trial 2 (S^S) and the response on Trial 2 (R_A) is followed by large reward. The SL animal thus learns S^S-R_A. Applying the same principles, it can be seen that the LS animal must learn S^L-R_A. While even in this simple example there are other aspects to the issue of what is learned, we are content here merely to indicate that the SL and LS animals did not learn the same things (to be frustrated, to accumulate extra attractions, etc.), but different things, i.e., S^S-R_A vs. S^L-R_A. This is evident from performance in extinction. In Leonard's investigation, Group SL was much more resistant to extinction than Group LS. Indeed, in comparison to a group trained on an LL schedule (large reward on Trials 1 and 2), Group SL showed increased resistance to extinction while Group LS did not, showing decreased resistance to extinction instead. In view of the fast extinction of Group LS, do we conclude that it did not accumulate extra attractions or that it was not frustrated? Questions of this sort are instructive. The essential condition for generating frustration or extra attractions is merely that the animal receive a smaller reward than expected. Why this should have been true for Group SL (increased resistance to extinction) and should not have been true for Group LS (no increase in resistance to extinction) raises considerable complications for those theories of frustration and dissonance which assume that internal stimuli can only be changed gradually, complications which, in our view, have not been fully faced by these theories.

At this point let us consider a slightly more complicated example of sequential effects in which the schedule of reward is varied while the magnitude of reinforcement is constant on all rewarded trials. Capaldi and Minkoff (1967), employing a minimum 20-minute intertrial interval (ITI), trained two groups under different reward schedules, one under a single alternating (SA) schedule, the other under an irregular (I) schedule. Both groups received the same number of rewarded (R) and nonrewarded (N) trials. In the present example we shall ignore what was learned following rewarded trials (in the presence of S^R), focusing instead on what was learned following nonrewarded trials (in the presence of S^N). The specific schedule received by each group on one of the acquisition days is shown in Table I.

In the SA group the animal learns that a single nonrewarded trial (symbolized as S^{N1}) is followed by reward, i.e., S^{N1}-R_A. Note, however, that in Group I reward follows not only a single nonrewarded trial (Trial 2–3) but four consecutive nonrewarded trials as well (Trial 8–9). The sequential model assumes

Table I

Reward Schedules Received by Group SA on All Days and by Group I on One of the Training Days in the Capaldi and Minkoff Investigation

	Trial									
	1	2	3	4	5	6	7	8	9	10
Group SA	R	N	R	N	R	N	R	N	R	N
Group I	R	N	R	R	N	N	N	N	R	R

that the memory for nonreward is modified as a function of successive nonrewards. This is symbolized as S^{N1}, S^{N2}, S^{N3}, S^{N4}, etc. Group I then learns two things, $S^{N1}-R_A$ and $S^{N4}-R_A$. The number of nonrewards in succession defines a variable called N-length. All other things being equal, the longer the N-length, the greater is resistance to extinction (e.g., Capaldi, 1966). In the Capaldi and Minkoff investigation, Group I, the longer N-length group, was more resistant to extinction than Group SA, the shorter N-length group, despite the fact that both groups received the same number of rewards and nonrewards. Many other investigations, some of them employing even longer ITI's than the minimum 20-minute ITI employed by Capaldi and Minkoff, could be cited as supporting the same point, that the organism's performance in acquisition or extinction or both is regulated by the sequence of N and R trials and is independent of the number of N and R trials (or the number of S and L trials, etc.). There are two major ways that such potentially embarrassing sequential data have been dealt with in the context of gradual change hypotheses.

V. Objections to the Sequential Analysis

One way is to suggest, for one reason or another, that the sequential data are artifactual. Surridge and Amsel (1965, 1966), for example, have suggested that differential external stimuli (E-produced cues, etc.) may accompany N and R trials. This view is not supported by experiments which have attempted to eliminate such cues directly (Bloom & Malone, 1968; Hanford & Zimmerman, 1969; Harris & Thomas, 1966). In a similar vein, Lawrence and Festinger (1962) have suggested that when the reward schedule is regular, the animal may learn to anticipate N trials and that anticipated N trials are not really "nonrewarding." It is interesting to note that by postulating such appropriate and sensitive "anticipation" Lawrence and Festinger are accepting a fundamental assumption of the sequential hypothesis, i.e., rapid internal stimulus change, while at the same time rejecting what they ordinarily assume, that expectation is increased or decreased slowly over repeated trials. Actually, Lawrence and Festinger's "anticipation notion" is incompatible with a great deal of data. For example, some studies have used too few trials to allow an anticipation to develop, yet have produced expected and predicted sequential effects (e.g., Capaldi & Hart, 1962; Capaldi & Lynch, 1968). Other studies have varied N-length in the context of schedules too irregular to be learned (e.g., N-length varies between 1 and 4 vs. 1 and 16) yet have produced expected sequential effects (e.g., Capaldi, 1964; Gonzalez, Bainbridge, & Bitterman, 1966; Gonzalez & Bitterman, 1964). The first way sequential data have been dealt with, then, is to suggest that they are artifactual.

We have already seen one example of the second way in which the implications of sequential data can in effect be minimized in connection with an interpretation supplied by Black and Spence, who postulated two sorts of PRE, one for "small trials," the other for "extended trials." The second way then, unlike the first, suggests that sequential data are valid, can be taken at face value, but only under specified restricted circumstances. In this connection, Amsel (1967), as well as others, has suggested that sequential variables may be effective, but only under massed trials. This view also suggests that there may be two sorts of PRE, one for massed trials (sequential variables), the other for spaced trials (frustration). The line of argument adopted by Black and Spence and by Amsel and by others is, of course, always permissible. It should be recognized, however, that it is either the strongest or the weakest line of argument a theorist can offer. If the theorist turns out to be right, it is the strongest line of argument because, in effect, no modification of his theory is required to deal with the recalcitrant data. If the theorist turns out to be wrong, however, it is the weakest line of argument because, in effect, it has been confessed in advance that in terms of his theory the data in question make little or no sense. It should be indicated that Amsel's dichotomy, and perhaps that of Black and Spence as well, is accepted by many. This is because many find it quite difficult to believe that the rat can remember whether or not it was rewarded 24 hours ago.

PART II

I. Experiment 1: Long ITI's, Irregular Schedules, and Magnitude of Reward

As we have seen, sequential variables and thus the memory mechanism are not considered to be operative according to some critics of the sequential hypothesis either when trials are widely spaced or when the reward schedule is irregular. In Experiment 1 we attempted to provide evidence relevant to these speculations. What is particularly interesting about Experiment 1, however is that by investigating these matters in the context of "reward magnitude" it was possible to test not merely the sequential view, but the frustration and dissonance hypotheses as well. This situation is rare. More typically, evidence which supports a sequential position either has no interpretation within these other contexts or is seen within these frameworks as resulting from experimental artifacts. Neither of these considerations seems to apply here; if the experimental outcome supports a sequential position, it appears to disconfirm the frustration and dissonance theories and vice versa.

According to the frustration and dissonance hypotheses, resistance to extinction is an increasing function of magnitude of partial reward. The larger the reward magnitude the larger respectively is r_g or expectation. Thus nonreward is more frustrating or dissonance-producing when accompanied by large reward. The frustration-dissonance view of magnitude of partial reward stems directly from experiments like those carried out by Hulse (1958) and Wagner (1961). In these, the ITI was 24 hours, the reward schedule was irregular and reward magnitude was either large (e.g., NLNNNLL, etc.) or small (e.g., NSNNNSS).

Note that in the Hulse-Wagner sort of investigation N trials are followed either by L (large reward) or S (small reward). A rewarded trial occurring after a nonrewarded trial defines an N–R transition. According to the sequential hypothesis, resistance to extinction is an increasing function of the reward magnitude contained in the N–R transition, i.e., N–L transitions should produce greater resistance to extinction than N–S transitions. If this is so, resistance to extinction is totally independent of "total magnitude of partial reward" and thus is more or less independent of processes such as amount of frustration or dissonance and so on. Much evidence has recently been presented (Capaldi & Lynch, 1968; Capaldi & Minkoff, 1971; Leonard, 1969) which simultaneously supports the sequential viewpoint on reward magnitude and disconfirms the frustration and dissonance hypotheses. Unfortunately, all of it has been collected under conditions of relatively massed trials and relatively regular schedules of reward. Thus, Capaldi and Lynch, employing about a five-minute ITI, showed that rats trained three trials each day under the regular schedule LLN were *less* resistant to extinction than rats trained SNS—Group LLN may be considered to have "zero" reward magnitude in the N–R transition, while Group SNS has a small magnitude of reward in the N–R transition. Employing even greater massing of trials than Capaldi and Lynch, Leonard also showed that a group trained SNL was much more resistant to extinction than a group trained LNS (total reward magnitude equated, magnitude of reward in the N–R transition varied).

Consider the sequential interpretation of these effects. If the transition is N–L, on the large reward trial the animal will remember having been nonrewarded (S^N) on the previous trial. Thus S^N will acquire control over the approach reaction, R_A, by means of large reward (S^N–R_A). If the sequence is N–S, however, S^N will acquire control over R_A by means of small reward (S^N–R_A). We assume that S^N acquires a greater capacity to control R_A when such control is established by means of large reward than by means of small reward (e.g., Capaldi, 1966). Thus in extinction (constant nonreward) a group previously trained N–L will run faster in the presence of S^N than a group previously trained N–S. In a group previously trained without an N–R transition (e.g., LLN), S^N does not directly acquire any capacity to control R_A and thus extinction should be very rapid. It can be seen that the available evidence in the area of magnitude of partial reward is consistent with the N–R transition hypothesis.

If we take seriously Amsel's notions on the effect of ITI, and Lawrence and Festinger's notions on the effects of regular schedules, we would tend to regard the agreement of the Hulse-Wagner findings with the predictions of sequential theory as a "fortuitous" occurrence. Is it? To determine whether it is or not we require an experiment which separates the effects, as the Hulse-Wagner experiments did not, of *total reward magnitude and magnitude of reward contained in the N–R transition* when the ITI is 24 hours and the reward schedule is irregular. Experiment 1, in addition to employing the groups of Hulse and Wagner (all hypotheses make the same predictions), employed two other groups in which total reward magnitude and magnitude of reward contained in the N–R transition were separated (hypotheses make different predictions).

A. METHOD

1. Subjects

The 36 male albino rats, obtained from the Holtzman Co., Madison, Wisconsin, were about 90 days old upon arrival at the laboratory. The Ss were randomly divided into four groups of nine Ss each.

2. Apparatus

The straight alley runways employed in the first three investigations reported here were of identical dimensions. The alley was 82 inches long, 4 inches wide, enclosed by 9 inch high sides, and covered with hinged ½-inch hardware cloth. It was constructed of wood and painted a midgray. It had three basic sections: *start* 14 inches, *middle* or *run* 52 inches, and *goal* 16 inches. When a 10-inch start treadle was depressed by the rat, whose front paws were always placed on the treadle's extreme forward edge, a .01-second clock started. This clock stopped and a second started when S broke an infrared beam 4 inches from the treadle's tip (*start time*). Interrupting the second beam, 52 inches from the first, stopped the second clock (*run time*) and started a third clock. Twelve inches from the second beam was a third infrared beam, located 2 inches from the front edge of a brass 2 x 4¼ x 1½ inch food cup which was covered by a tightly fitting, automatically controlled, sliding metal lid operated by an electric motor. Interruption of the third beam stopped the third and last clock (*goal time*) and opened the lid covering the food cup. When S broke the third beam, a brass guillotine door 12 inches from the alley's distal end was lowered manually. The elapsed time on the three clocks was summed and is termed *total time.*

3. Preliminary Training

Upon arrival at the laboratory the rats were housed in group cages and given free access to food and water for 5 days. On the sixth day (Day 1 of pretraining) all *S*s were placed in individual cages and restricted to a 12 gm per day maintenance diet of Wayne Lab Blox with *ad lib.* water. All animals were handled daily during pretraining and were habituated to the unbaited apparatus in groups of 4 for 5 minutes on Days 7 and 9. On Days 7 through 10 of pretraining each rat received ten 0.045 gm Noyes pellets in the home cage.

4. Experimental Training

The 22 days of acquisition training were administered to each group on the basis of the schedules shown below:

```
Group LNL: L  L  L  N  N  N  L  N  N  L  N  N  N  L  L  L  L  L  N  N  L  L
Group SNS: S  S  S  N  N  N  S  N  N  S  N  N  N  S  S  S  S  S  N  N  S  S
Group SNL: S  S  S  N  N  N  L  N  N  L  N  N  N  L  S  S  S  L  N  N  L  L
Group LNS: L  L  L  N  N  N  S  N  N  S  N  N  N  S  S  L  L  L  N  N  S  S
```

Groups LNL and SNS correspond to the sorts of groups employed by Hulse and Wagner, i.e., total magnitude of reward and reward magnitude contained in the N–R transition are confounded. Note that all groups have four N–R transitions. Groups SNL and LNS allow a test of the hypothesis that differences in the Hulse-Wagner situation were not related to total magnitude of reward per se, but were regulated, in fact, by magnitude of reward contained in the N–R transition, i.e., N–L vs. N–S. Note that over days, Groups SNL and LNS receive the same total magnitude of reward. However, Group SNL like Group LNL (which receives overall greater total magnitude of reward) receives N–L transitions, while Group LNS like Group SNS (which receives lesser total magnitude of reward) receives N–S transitions.

Acquisition training began following the tenth day of pretraining. There was one trial per day in acquisition and extinction (26 days). Reward magnitude was either large (L), twenty-two 0.045 gm Noyes pellets, or small (S), two pellets. *S* was never removed from the goal box until all pellets had been consumed, and then was removed immediately. On all nonrewarded (N) trials, *S* was confined to the goal box for 30 seconds.

The *S*s were run in squads of four, one *S* from each group in each squad. The within-squad running order was varied daily. On all trials, *S* was placed in the apparatus as described above, and given 60 seconds to traverse each alley section. If *S* failed in this, it was assigned a time score of 60 seconds for that alley section and all not yet traversed alley sections, and was picked up and placed in the goal box. Following a trial, the animal was returned to the home cage, where 15 minutes later it received the 12 gm daily ration minus the amount eaten in the apparatus.

B. RESULTS

1. Acquisition

Running speeds (total speeds) for each of the four groups on each day of acquisition are shown in Fig. 1. The most obvious feature to be noted in Fig. 1 is that on the final days of acquisition the three groups which received a large magnitude of reward were running equally rapidly and that all of them were running faster than Group SNS, which received only a small magnitude of reward. An analysis of total speeds on the last day of acquisition indicated that the differences between Group SNS and the other three groups were highly significant $(F = 10.18, df = 3/32, p < .001)$. A second, and perhaps more important feature of the running speeds shown in Fig. 1 is that none of the groups showed any tendency to anticipate either rewarded trials or nonrewarded trials. The reader may ascertain this for himself by comparing running speeds on a particular trial with the reward condition actually occurring on that trial. It will be seen, first of all, that the dominant tendency early in training was to run relatively fast following rewarded trials and relatively slowly following nonrewarded trials. This tendency sometimes produced "errors" from an anticipatory standpoint, in that speeds were sometimes higher on nonrewarded than on rewarded trials. For example, on Trial 4, a nonrewarded trial (which followed three rewarded trials), speeds were higher than on Trial 7, a rewarded trial (which followed three nonrewarded trials). Later on in training, after the animals had more or less overcome the tendency to slow down following nonrewarded trials, i.e., from about Trial 14 onward, speeds appear to be the same on rewarded and nonrewarded trials. Of course, given the irregularity of the schedules employed, it perhaps could not be seriously expected that the animals would learn to anticipate reward or nonreward. However, in view of the importance attached to this matter by Lawrence and Festinger, it seems

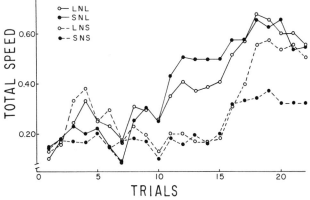

Fig. 1. Speed of running (total speeds) in each of the four groups on each day of acquisition (Experiment 1).

advisable to indicate that our *a priori* expectations are consistent with the acquisition data which were actually obtained.

Because of the terminal acquisition differences, the speed scores were transformed into rate measures as recommended by Anderson (1963). The speeds on the last 6 days of acquisition and on the first day of extinction were used to estimate the asymptote of acquisition, while the extinction asymptote was taken to be the reciprocal of 60 seconds in each alley section and 180 seconds in total.

2. Extinction

The rate measures for *total* speed for each of the four groups on each day of extinction are shown in Fig. 2. Groups LNL and SNL differed from each other to a negligible degree and showed greater resistance to extinction than Groups LNS and SNS. The rate measures for Group LNS tended to fall below those of Group SNS from about Trial 10 onward. The difference in resistance to extinction of Group LNS relative to Group SNS was largest when measured in terms of *run time*, where it approached but did not reach significance and reflects differences in acquisition asymptote. On the basis of the speed scores themselves, Groups SNS and LNS were quite comparable following the initial extinction trials.

A simple analysis of variance employing the rate measures over all of the extinction trials indicated that differences for *total time* were significant ($F = 11.87$, $df = 3/32$, $p < .001$), as they were in the run and goal sections, but not the start section. A subsequent Duncan's range test indicated that neither in *total time* nor in time taken to traverse any alley section did Group LNL differ

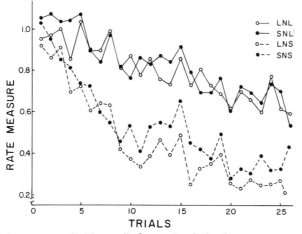

Fig. 2. Rate measures (total speed) for each of the four groups on each day of extinction (Experiment 1).

from Group SNL. Similarly, Groups SNS and LNS failed to differ, except, as previously mentioned, in *run time*, where differences approached but did not reach significance ($.05 < p < .10$). In *total, run* and *goal times*, but not in start time, Groups LNL and SNL differed significantly from Groups SNS and LNS ($p < .01$).

C. DISCUSSION

Experiment 1 indicated that when the ITI is 24 hours and when the reward schedule is irregular, resistance to extinction is nevertheless independent of total reward magnitude and is an increasing function of reward magnitude contained in the N–R transition. The results of Experiment 1 are consistent with all previous partial reward magnitude investigations including those of Hulse and Wagner and those employing relatively massed trials and relatively regular schedules. Experiment 1 suggests that the N–R transition is responsible for the partial reward magnitude effect and that the memory of previous nonreward, S^N, acquires the capacity to control the instrumental reaction, R_A, even when trials are widely spaced and the reward schedule is irregular. Accordingly, Experiment 1 suggests that S^N is produced by means of a memory mechanism as opposed to a trace mechanism (Sheffield, 1949) or a learning mechanism (Amsel, 1958; Spence, 1960).

The results of Experiment 1 are significant not only for what they support, but also for what they fail to support. For example, neither the frustration hypothesis nor the dissonance hypothesis appear capable of dealing with the results of Experiment 1. This shortcoming is serious because the results of Experiment 1 are in no way peculiar, but seem of a piece with all previous findings on magnitude of partial reward. An inability to deal with the results of Experiment 1 constitutes, in effect, an inability to deal with magnitude of partial reward per se. Indeed, it is possible to generalize even further. It has long been known (e.g., Grosslight *et al.,* 1953) that in the absence of N–R transitions resistance to extinction will not be increased by partial reward. If, as previously indicated, we conceptualize an absence of N–R transitions as "zero" reward magnitude, then findings of the kind reported by Grosslight *et al.* are obviously continuous with the results of Experiment 1. In other words, magnitude findings (Experiment 1) and schedule findings (see Experiment 2) are intimately related. In this context, the failure of theories of the frustration and dissonance type consists not merely in failing to understand reward magnitude specifically, but in failing to understand the nature of the variables and mechanisms underlying partial reward generally. Our sole reason for emphasizing these points is to make clear that, in our view at least, no clearly viable alternative exists at this time to understand partial reward magnitude effects outside of sequential theory. While one may nevertheless dismiss a memory mechanism, one cannot do so lightly.

II. Experiment 2: N-Length vs. Number of Nonrewards

There is a broad literature dealing with the effects of the percentage of reward and schedule of reward. Schedule of reward is defined here as the specific sequence in which rewarded and nonrewarded trials occur, e.g., NRNR vs. RNNR. When percentage is varied, schedule necessarily varies, although most investigators of percentage ignore this fact. Deliberate attempts to vary schedule (e.g., Gonzalez *et al.,* 1966) may involve holding percentage of reward constant or allowing it to vary. As we see matters, the theoretical importance of the percentage-schedule literature cannot be overestimated. Unfortunately, the implications of this literature do not appear to be widely understood, perhaps because it is so complicated. In any event, to us it seems probable that theories based on gradual internal stimulus change may be in principle unable to explain the percentage-schedule literature. On the other hand, the percentage-schedule literature seems easily understandable on the basis of a memory mechanism and supports such a mechanism. For these reasons, then, the percentage-schedule literature would seem to be of the utmost importance to those of us interested in memory.

Variables usually equated in level are systematically varied in Experiment 2, thus generating a unique experimental design. Because of this, Experiment 2 can only be understood in the context of past research in the percentage-schedule area. The primary purpose of Experiment 2 is to assess the relative importance of N-length vs. number of nonrewards on resistance to extinction and to determine if the number of N-trials has any effect whatever on resistance to extinction. The variable N-length is defined in terms of the number of nonrewarded trials which precede a rewarded trial, i.e., N-length may be "one" (NR), "two" (NNR), "three" (NNNR), and so on. A given N-length may occur once (NR), twice (NRNR), and so on. The major previous experimental findings forming the context for Experiment 2 are reviewed in Section II,A, being divided into four "cases."

A. SUMMARY OF PERCENTAGE-SCHEDULE FINDINGS TO DATE

1. Case 1

In this work, percentage of reward is varied while number of acquisition trials is held constant. The effects of percentage of reward on extinction depend upon level of acquisition training (Bacon, 1962). Early in training, resistance to extinction increases as the percentage of reward increases and later in training it increases as percentage decreases. Few theorists, if any, consider that these effects are due to percentage of reward per se. Lawrence and Festinger (1962) and Weinstock (1954, 1958) suggest that resistance to extinction increases as number of nonrewarded trials increased. (In Case 1, as the percentage of reward decreases, the number of nonrewarded trials increases.) While this view is consistent with the late-trial percentage effect, it is not consistent with the early-trial effect. Frustration theory has yet to specify systematically what the role of any particular variable may be in the percentage-schedule area. According to Capaldi (1966), the early-trial percentage effect is regulated by the number of times a particular N-length occurs (see Spivey, 1967), while the late-trial percentage effect is related, perhaps exclusively, to N-length. In this connection it should be noted that investigations of the effects of percentage of reward typically assign rewarded and nonrewarded trials in a quasi-random fashion which results in the lower percentage groups receiving longer N-lengths.

2. Case 2

Under some experimental conditions, if the number of nonrewards is equated, but the percentage of reward varies, resistance to extinction does not differ (Lawrence & Festinger, 1962; Uhl & Young, 1967). In cases of this sort, the several groups received the same number of nonrewarded trials, e.g., 10, but different numbers of rewarded trials, e.g., 10 vs. 20 (10 R–10 N vs. 20 R–10 N). Equal resistance to extinction in such groups is taken as support for the dissonance hypothesis. Unfortunately, investigations employing this case have not described the specific reward schedules they have employed. That the results obtained may have been due to the schedule of reward rather than to the number of nonrewards is strongly suggested by the third and fourth classes of experiments described next.

3. Case 3

In investigations of this type, the number of rewards and nonrewards is equated while the sequence of rewards is varied, e.g., NRNR vs. RNNR. If a small number of acquisition trials is employed, resistance to extinction is found to increase as the number of transitions from N to R increases. For example, Group NRNR would be more resistant to extinction than Group RNNR (cf. Capaldi, 1964; Capaldi & Hart, 1962; Capaldi & Wargo, 1963). This corresponds exactly to what occurs when percentage of reward is varied and the number of acquisition trials is small (Case 1, Spivey, 1967), strongly suggesting that the two cases are theoretically identical. If considerable acquisition training is given, however, resistance to extinction increases as N-length increases (e.g., Capaldi, 1964; Gonzalez et al., 1966; Gonzalez & Bitterman, 1964); that is to say, later in training, Group RNNR would be more resistant to extinction than Group NRNR, again corresponding to Case 1. In the present case, then, resistance to extinction varies when the number of nonrewards is constant. This is not necessarily in conflict with the findings of Case 2 described above (10 R–10 N vs. 20 R–10 N), because there are many schedules which can be selected which will result in equal resistance to extinction in groups given the same number of nonrewards.

4. Case 4

Indeed, one can deliberately create a situation in which a group trained on a schedule delivering a higher percentage of rewards receives longer N-lengths than a lower percentage group. When this is done and considerable acquisition training is employed, the higher percentage group (fewer number of nonrewards) is actually the more resistant to extinction (Capaldi & Stanley, 1965). This finding, as well as those considered under Case 3, is clearly incompatible with the conclusion that the Case 2 findings are due exclusively to the number of nonrewards. It should be noted for later reference that the groups employed by Capaldi and Stanley received essentially "irregular" or "unpredictable" schedules of reward.

Now it is possible, as Amsel (1967) and Lawrence and Festinger (1962) recommend, to fragment this literature by looking for differences among investigations on the basis of long vs. short ITI's or regular vs. irregular schedules and so on. Sequential theory does not do that. On the contrary it suggests that the percentage-schedule literature should be dealt with as a whole. This literature suggests, especially for Cases 3 and 4, that by a suitable arrangement of schedule of reward it is possible to obtain differences in resistance to extinction either when the number of nonrewards is held constant or when a group actually receives a smaller number of nonrewards. *Note that this may suggest only that sequence of reward is a more powerful determinant of resistance to extinction than number of nonrewards. It does not necessarily suggest that the number of nonrewards does not determine resistance to extinction.* From a sequential standpoint, in order to determine what influence, if any, the number of nonrewards may have, groups would have to be equated, say, for N-length while receiving different numbers of nonrewards. An example of this is shown below:

RNR   vs.   RNRN

While both groups receive an N-length of 1 (Trial 2–3), Group RNRN receives an additional nonrewarded trial (Trial 4). If the number of nonrewards per se has any effect on resistance to extinction outside of their inclusion in an N-length, then Group RNRN should be more resistant to extinction than Group RNR. In order to round out a design of this sort, we would obviously want to include a group trained RNNR. We could then evaluate the influence of a nonrewarded trial when contained in a sequence of nonrewarded trials, as opposed to the effect which such trials have when they are isolated from one another by rewarded trials (RNNR vs. RNRN) as well as evaluate whether or not the number of nonrewards per se influences resistance to extinction (RNRN vs. RNR). Experiment 2 contained the five groups shown in schematic form below:

1. RNR
2. RNRN   4. RNNR
3. RNRNN   5. RNNNR

There are two additional features of Experiment 2 worth mentioning. First, the minimum ITI between trials never fell below 20 minutes. Secondly, on different days the group designated above as RNR was trained either RRNR or RNRR, a similar procedure being followed in the case of the remaining groups (e.g., RRNRN or RNRRN). This was done so that the Ss would not learn that R was always followed by N, thus forming a discrimination.

B. METHOD

1. Subjects

The Ss were 60 naive male rats about 90 days of age when purchased from the Holtzman Co., Madison, Wisconsin.

2. Apparatus

The straight alley runway employed was of the same description as that used in Experiment 1.

3. Procedure

On arrival at the laboratory, Ss were placed in individual cages and for 5 days given *ad lib.* food and water. On Day 6 each S was randomly assigned to one of three replications, each consisting of 20 Ss. The first replication began deprivation on Day 6, the second on Day 11, and the third on Day 17. The Ss not beginning deprivation on Day 6 continued on *ad lib.* feeding until their replication began its deprivation schedule. Deprivation consisted of a 12-gm maintenance schedule, which continued throughout the experiment, water being given *ad lib.* The Ss were handled outside the home cage for approximately 5 minutes in groups of five on Days 13, 14, and 15 of deprivation. On handling days each S was fed eight 0.045-gm Noyes pellets in the home cage. On Day 16 of deprivation the Ss in groups of five were allowed to explore the alley for a period of 10 minutes. On Day 17 each S was twice placed placed in the goal box and permitted to consume the eight pellets in the goal cup. The guillotine door was lowered before S was placed in the goal box on the first of these trials, whereas on the second placement the door was lowered as the S was being put into the goal box. The placements were separated by 20 minutes. Each replication consisted of four Ss from each group.

Each group received three rewarded trials daily but different numbers of N trials. On odd days, Groups RNR, RNNR and RNNNR were trained with the trial sequence RRNR, RRNNR, and RRNNNR, respectively. On even days, the groups were trained RNRR, RNNRR, and RNNNRR. On odd days, Groups RNRN and RNRNN were trained RRNRN and RRNRNN, while on even days they were trained RNRRN and RNRRNN. The 10 days of acquisition training were followed by 7 days of extinction training at six trials each day. On Day 1 of extinction the initial two trials were rewarded.

The *S*s within a replication were run in rotation with a minimum intertrial interval of 20 minutes with the following restrictions: Half of the *S*s having only four trials per day were run on rotations 1–4 and the remaining on rotations 3–6. *S*s with five trials per day were also split, half not being run on the first rotation and the remaining not running on the sixth rotation. The splits were divided equally within groups. The order of running was changed daily so that an *S* which ran first on one day ran last on the next, while all the members of the replication moved up one position in the running order.

On all rewarded trials *S* was removed immediately after having eaten the twelve 0.045-gm Noyes pellets in the food cup. On all nonrewarded trials goal box confinement lasted for 30 seconds. If an *S* failed to traverse the runway in less than 180 seconds, it was placed in the goal box. The maximum time recorded for any one section was 60 seconds.

C. Results

Figure 3 presents total speeds for each of the five groups on each day of acquisition and extinction. An analysis over the last 3 days of acquisition indicated that differences between the groups were not significant in any alley section or in total running speed (all values of $F < 1$). Subsequently, t tests were performed to determine whether Group RNRN or Group RNRNN were discriminating between R trials and N trials. The t test compared difference

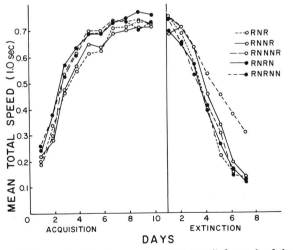

Fig. 3. Acquisition and extinction speeds (total speed) for each of the five groups on each day of acquisition and on each day of extinction (Experiment 2).

scores between each S's mean R trial speed and mean N trial speed on the last day of acquisition. The obtained values of t of 1.32 for Group RNRN and 1.09 for Group RNRNN were not significant ($df = 11$ in each case). The analyses reported above then suggest that the terminal acquisition speeds of the groups did not differ and that neither Group RNRN nor Group RNRNN were discriminating between R trials and N trials.

It can be seen in Fig. 3 that Group RNNNR was clearly the most resistant to extinction and that Group RNNR was slightly more resistant to extinction than the remaining groups. An analysis performed over the extinction speeds shown in Fig. 3 indicated that overall differences between the groups were significant ($F = 4.55$, $df = 4/55$, $p < .01$). A subsequent Duncan's range test indicated that Group RNNNR differed significantly from Group RNNR ($p < .05$) and from all of the remaining groups ($p < .005$) with no other differences being significant.

D. DISCUSSION

Experiment 2 suggests that the number of nonrewards per se has no effect whatever on resistance to extinction. Apparently, if nonreward is to increase resistance to extinction, it must be contained within an N-length. This effect, be it noted, was demonstrated at a minimum ITI of 20 minutes. Group RNNR (N-length = 2) was only slightly more resistant to extinction than the various groups in which the N-length was unity. Group RNNNR (N-length = 3), on the other hand, was much more resistant to extinction than Group RNR, Group RNRN or Group RNRNN. Within the sequential framework it is assumed that $S^N 1$ (produced by a single nonreward), $S^N 2$ (produced by two successive nonrewards), $S^N 3$, etc., are arrayed along a continuum of approximately logarithmic scale, that is to say, differences between the successive memories $S^N 1$, $S^N 2$, $S^N 3$, etc. decrease. This implies, as was found here, that when successive N-lengths are compared, e.g., 1 vs. 2, 2 vs. 3, etc., differences in extinction will increase initially, e.g., the difference 2 vs. 3 would be greater than the difference 1 vs. 2, and then decrease. Let us, however, avoid these matters of detail, turning instead to the more general implications of these findings.

We stated earlier that the percentage-schedule area strongly implicates the presence of a mechanism of rapid internal stimulus change in the rat and thus of a memory mechanism, and for this reason is critically important in a systematic sense. At an intuitive level, at least, the literature review provided earlier and the results of Experiment 2 perhaps serve to demonstrate these points. More systematic, that is to say explicitly theoretical, treatments are provided elsewhere (e.g., Capaldi, 1966, 1967). It was suggested earlier that views based on gradual internal stimulus change may in principle be inadequate to deal with percentage-schedule findings. If this is so, it would tend to increase our confidence in a memory approach to learning phenomena. While we cannot, in the limited space available, demonstrate as well as we would like all of the difficulties faced by gradual stimulus change approaches in relation to percentage-schedule findings (and thus by implication, to other learning phenomena where the gradual change approach seems to work better), we can, we think, at least illustrate strongly the general nature of these difficulties. While the frustration hypothesis, which is considered to be a theory of partial reward, has attempted to explain why partial reward produces greater resistance to extinction than consistent reward (e.g., Amsel, 1958), it has not, as yet, in even a single instance attempted to explain in a systematic fashion differences among groups which have all received partial reward but with differences in percentage or schedule. This surprising observation in itself conveys a sense of the difficulties faced by gradual stimulus change hypotheses when confronted with percentage-schedule data; after all, the frustration hypothesis has been around for over a decade. It is all well and good to say, as so many have, that resistance to extinction increases as percentage decreases because as percentage decreases frustration increases. However, it is an entirely different matter to demonstrate from what combination of theoretical principles and identified experimental variables this deduction flows, and having done that, to determine how these principles and variables fare when applied to other experimental data in the percentage-schedule area. While one cannot judge the adequacy of a theoretical solution he has not seen, he can discern the difficulties that solution, whatever it may be, will face. Generally speaking, given the structure of the frustration hypothesis, it seems very likely that it will have to fractionate the percentage-schedule literature into classes of various sorts. Some classes of data will be relevant to frustration, others will not be relevant, and so on. This is because a hypothesis of the frustration variety presumably identifies classes of variables (number of nonrewards, ratio of rewarded to nonrewarded trials, etc.) which cannot possibly work when applied across the entire percentage-schedule area (e.g., see Cases 3 and 4 described earlier). This judgment, of course, suffers from a certain ambiguity inasmuch as frustration has not as yet systematically identified any variable as relevant in the percentage-schedule area. In any case, even in the absence of such identification, it is clear that such fractionation as described above is contemplated. Thus Amsel (1967) proposes to deal only with partial reward data which have been collected at long ITI's. About the only issue that can be joined at this time, then, is how reasonable or unreasonable it is to suggest that the percentage-schedule area be fractionated on the basis of ITI. Experiments 1 (24-hour ITI) and 2 (20-minute ITI) reported here suggest that it is not at all reasonable. Experiments 1 and 2 aside, it is clear enough that percentage-schedule data collected at relatively short ITI's (for example by Bacon, 1962) and at long ITI's (for example, by Weinstock, 1954, 1958) are quite similar, and thus may be assumed to be regulated by identical variables and processes.

Our second example of a gradual internal stimulus change view is the dissonance hypothesis. Unlike the frustration hypothesis, dissonance theory has attempted to deal with percentage-schedule data. It suggests, as previously indicated, that resistance to extinction should increase as number of nonrewards increases ($R_E f N$). As we have seen, numerous experimental results, including those of Experiment 2, are inconsistent with the generalization that $R_E f N$. In any event, since most of the comments raised in connection with the frustration hypothesis apply with slight modification to dissonance theory, we would add only the following observation, which suggests perhaps more directly the sorts of problems gradual internal stimulus change hypotheses face in relation to percentage-schedule data. It is difficult to understand how, within the context of their overall theory, Lawrence and Festinger could have arrived at the conclusion that $R_E f N$, since this conclusion *does not seem consistent with the dissonance hypothesis.* The dissonance hypothesis, like the frustration hypothesis, assumes that what happens on nonrewarded trials (dissonance, frustration) is bigger or more intense the greater the expectation of reward, and the bigger or more intense this dissonance or frustration is, the greater should be the resistance to extinction. One variable controlling expectancy according to Lawrence and Festinger is the number of rewarded trials. Should not, then, expectancy and resistance to extinction be greater for a group trained 20 R–10 N than for a group trained 10 R–10 N inasmuch as the former group receives twice as many rewarded trials as the latter group? It would seem so. Of course it may not be so, since other factors may enter, and so on. Unfortunately, Lawrence and Festinger do not explicitly identify the steps in their reasoning whereby they conclude that $R_E f N$, and we can therefore do little more here than note the following: First, we do not know what role is assigned to expectancy or other factors by Lawrence and Festinger in deducing within a dissonance framework that $R_E f N$. Second, we do not even know if the deduction $R_E f N$ can be made consistent within a dissonance framework, if expectaction is assigned the role it is normally assigned in the overall system. Generalizing from this case, we would say that gradual change hypotheses will always encounter ambiguities of this sort as they attempt to deal with percentage-schedule data.

Actually, there is one circumstance in which we are prepared, at this time, to say that the proposition that $R_E f N$ can be defended within a dissonance framework. The conclusion $R_E f N$ can clearly be defended in that context if it is assumed that expectation grows rapidly, perhaps even reaching a maximal value following a single rewarded trial. If this were so, then Groups 10 R–10 N and 20 R–10 N would have the same expectation level and therefore would experience the same amount of dissonance over their 10 nonrewarded trials. Note that the notion that expectancy level reaches a maximal value following a single rewarded trial is explicitly assumed in the memory model. In the discussion of Experiment 3, we shall see that Amsel, Hug, and Surridge (1968) have lately seen fit to adopt the notion of a very rapidly increasing r_g and r_f level.

III. Experiment 3: Small Trial PRE: N-Length without Prior Reward

The assumptions entertained by sequential theory, that internal reward-related stimuli are present from the very outset of training and that trial-to-trial variations in these stimuli are large or extensive, arose initially on the basis of acquisition data from extended trials of the kind presented in Experiment 2. These assumptions stem rather naturally from a memory approach and to some extent, perhaps, define what seems to be meant by memory. Critical and interesting support for these assumptions has been provided by the pioneering work of McCain (1966). McCain has demonstrated that a PRE will occur following a very small number of acquisition trials—perhaps as few as two trials (nonreward followed by reward, or NR). While McCain's innovative results have been confirmed and extended in other laboratories (e.g., Capaldi, Lanier, & Godbout, 1968a; Padilla, 1967), much remains to be learned about the conditions under which the small trial PRE will occur. Determining the nature of these conditions appears to have critical theoretical importance as we will show below.

We saw earlier in the case of the frustration hypothesis that, generally speaking, the animal's internal stimulus environment is assumed to lag considerably behind the prevailing conditions of reward. The difficulties created for hypotheses of the frustration type by the small trial PRE are clear; if it takes x trials to establish s_f-R_A and the PRE is shown to occur following fewer than x trials, then something is obviously wrong with the hypothesis. While hypotheses other than the frustration hypothesis are not supported by the small trial PRE (e.g., Lawrence & Festinger, 1962; Weinstock, 1954), only theorists within the frustration framework have attempted, as yet, to deal with it. Three suggestions have thus far been advanced. Black and Spence (1965), as we have already seen, ascribed the small trial PRE to sequential mechanisms, suggesting that frustration comes into play only later in training. They thus adhered to the original assumptions of frustration theory. Surridge, Rashotte, and Amsel (1967) failed to obtain a small trial PRE and suggested that previous positive results might be artifactual. Subsequently, Amsel et al. (1968) were able to obtain a small trial PRE, and in an attempt to deal with it, proposed what appears to be a radical modification of frustration theory. We are concerned here with testing their hypothesis.

According to Amsel et al., when the rat is rewarded with a single large pellet, it will eat the pellet without moving about the goal box. However, when rewarded with many small pellets (or wet mash), it will eat a few pellets, move about the goal box, return to eat a few more pellets and so on. The minimal conditions for the occurrence of the small trial PRE are the following: First, the animal must receive a rewarded trial in the apparatus in order to build up r_g—there is a suggestion that r_g develops faster when reward is given in the form of many small pellets, rather than as a single large pellet. Second, a nonrewarded trial must occur in order to generate frustration. Finally, a rewarded trial must occur. On this trial the animal moves about the goal box, as previously described, thus repeatedly conditioning the stimuli produced by anticipatory frustration, s_f to the approach reaction, R_A.

This elaborate model pushes the learning conception to a limit—r_g is now said to be built up on the basis of a single rewarded trial and on a subsequent nonrewarded trial not only is frustration generated, but anticipatory frustration (r_f) is conditioned to the cues of the apparatus on that nonrewarded trial $(S_A$–$r_f)$. Even when pushed to this limit, however, the learning model seems inadequate, perhaps even at the time it was published. First, McCain has demonstrated, as previously indicated, that a small trial PRE will occur following the two trial sequence NR, i.e., the small trial PRE occurred even when the N trial was not preceded by an R trial (no r_g, thus no r_f). Furthermore, Spear, Hill, and O'Sullivan (1965) and Spear and Spitzner (1967), in a series of experiments which controlled many variables, reported that a series of nonrewards (12 or 24) not preceded by reward, but, of course, followed by reward, produced a PRE, an effect they call the initial nonreward effect (INE). The INE was later obtained by Robbins, Chait, and Weinstock (1968). On the basis of their experiments, Spear and Spitzner conclude that the INE and the PRE are governed by identical processes. Amsel *et al.,* in presenting their small-trial PRE hypothesis, did not refer either to McCain's NR results or to the INE of Spear and co-workers. Perhaps this was because a series of nonrewards was employed in the INE investigations, and considerable prior alley exploration was employed in McCain's study. Thus none of these studies may have been thought of by Amsel *et al.* as "small trial PRE studies." In any event in the present investigation only a small number of nonrewarded trials was employed. First, unlike McCain's study, all preliminary adjustment to the apparatus was eliminated. Second, unlike the studies of Spear and his co-workers and of Robbins *et al.*, only a small number of "experimental" nonrewards were employed (2 or 5). Needless to say, these were not preceded by a rewarded trial in the apparatus. Indeed, stern experimental measures, similar to those employed by Spear and co-workers, were adopted. The rats were never fed outside the home cage, except on the experimental trials. Furthermore, when fed in the home cage, the animals were always fed by a gloved E but were removed from the home cage both in preliminary training and experimental training by a different ungloved E. The purpose of these procedures was not merely to prevent generalization of the hypothetical r_g from the home cage to the experimental apparatus, but to establish, if possible, a discrimination, i.e., "when I am removed from the home cage I am never fed."

A. Method

1. Subjects

The Ss were 78 male rats, 90 days old when obtained from the Holtzman Co., Madison, Wisconsin.

2. Apparatus

The runway employed was of identical dimensions to that used in Experiment 1.

3. Procedure

During the 21-day period preceding the experimental phase, Ss were frequently handled, adjusted to a 12-gm daily ration, and given experience eating 0.045-gm Noyes pellets in the home cage. The E who handled the Ss was ungloved. A second, gloved E fed the Ss. The Ss were never fed at handling, nor were they ever fed less than 30 minutes following handling.

There were four groups of Ss. Group N_S ($N = 26$) received two nonrewarded trials (N) followed by two rewarded trials (R) or NNRR (N-length = 2). Group N_L ($N = 26$) was trained NNNNNRR (N-length = 5). Group R_S ($N = 13$) received four rewarded trials while Group R_L ($N = 13$) received seven rewarded trials. All acquisition trials occurred on Day 22, the Ss being run in squads of six, one S from Groups R_S and R_L and two Ss from Groups N_S and N_L. The resulting ITI was about 6–8 minutes. On the day following acquisition training all Ss were given 10 extinction trials. If an S did not enter the goal box within 60 seconds, it was placed there by E and a time score of 60 seconds was recorded.

B. Results

Trial 1 of extinction is considered to be an acquisition trial. Figure 4 shows the running speeds of each of the four groups on Trial 1 of extinction, and on the nine succeeding extinction trials in blocks of three trials in each alley section. Analyses of Trial 1 of extinction indicated that in none of the alley sections were the differences significant (the largest F was 1.25 on the goal measure). It can be seen in Fig. 4 that as we advance from the start section to the goal section, there is an increasing tendency for the partial reward groups to extinguish more slowly than the consistent reward groups and for Group N_L, the longer N-length group, to show an increasing superiority in extinction relative to Group N_S the shorter N-length group. A trials-by-groups analysis, performed on the data from each alley section bears out statistically what can be seen in Fig. 4. In the start section, as can be seen, the groups were pretty much alike. In the run section the Trials × Schedule interaction was significant ($F = 2.63$, $df = 9/779$, $p < .01$) indicating that the partial groups extinguished at a slower rate than the consistent groups, and the Trials × N-length interaction approached significance

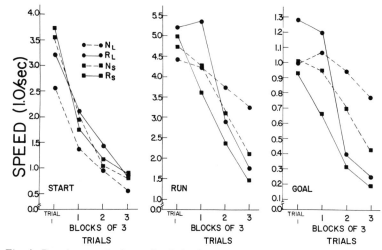

Fig. 4. Running speeds for each of the four groups in each alley section on the first extinction trial of the day and subsequently in blocks of three trials (Experiment 3).

($F = 1.64$, $df = 9/779$, $p<.10$), indicating that Group N_L was extinguishing at a slower rate than Group N_S. In the goal section the partial reward vs. consistent reward main effect was significant ($F = 5.87$, $df = 1/74$, $p < .025$), and the N-length main effect approached significance ($F = 2.75$, $df = 1/74$, $p <.10$). Moreover, the interaction of trials and reward schedule was significant ($F = 4.15$, $df = 9/779$, $p <.01$), as was the interaction of trials and N-length ($F = 2.03$, $df = 9/779$, $p < .05$).

C. DISCUSSION

Experiment 3 indicates that in order to obtain a small trial PRE, it is only necessary to follow nonreward by reward (N-R transition). This finding supports and lends generality to earlier findings of a similar kind reported by McCain, by Spear and his co-workers, and by Robbins *et al.* That is, the present investigation obtained the "INE" employing a minimal number of nonrewards, thus clearly identifying the INE as a small trial phenomenon. The present results, consistent with earlier ones (e.g., Godbout, Ziff, & Capaldi, 1968), indicate that the small trial PRE is larger in the goal section than in earlier alley sections. Furthermore, N-length was shown to be an effective variable under the present experimental conditions, adding further support to the conclusion that the small trial PRE and the extended trial PRE are regulated by the same variables and processes, a conclusion recommended by many (e.g., Capaldi *et al.*, 1968a; McCain, 1966; Robbins *et al.*, 1968; Spear & Spitzner, 1967). Spear and Spitzner found no reliable tendency toward an N-length effect (12 vs. 24), possibly because the smaller N-length of 12 was itself rather long.

The present investigation indicates that the minimal necessary conditions for obtaining a small trial PRE postulated by Amsel *et al.* are not minimal enough. Specifically, it is not necessary that the "N" trial be preceded by an "R" trial. In the absence of this "R" trial, it does not seem feasible to talk about the animal being frustrated on the "N" trial. Thus, even when one assumes the optimal conditions for the generation of r_g and r_f, the classical conditioning learning model breaks down. It would appear that no matter how manfully one tries to get the learning model to supply the necessary internal stimuli on a fast enough basis, it cannot be induced to do so. The learning model is too cumbersome; it does not appear to be relevant to the process whereby the animal is supplied with internal stimuli. The alternative recommended here is to assume that the internal reward-related stimuli are provided not according to the laws of learning but, as previously indicated, according to the laws of memory. The small trial PRE data, like schedule data, percentage data, and reward magnitude data considered earlier, are consistent with the memory approach.

It should perhaps be indicated, in the interest of completeness, that since the small trial PRE hypothesis was published by Amsel *et al.,* other data which are inconsistent with it have appeared. Thus Howlett and Sheldon (1968) and McCain (1969) have shown that a small trial PRE will occur when the reward is delivered in the form of a single large pellet; that is to say, many small pellets are not necessary. In another investigation (Godbout *et al.,* 1968) various groups were given either no goal box placements, rewarded goal box placements (r_g), or nonrewarded goal box placements (no r_g) prior to the running trials. Presumably, the rats placed in the baited goal box should have had the larger r_g and thus should have shown the larger PRE. This was not the case in terms of rate of extinction on any other extinction measure; the nonrewarded placed group showed a PRE as large as that of the rewarded placed group.

IV. Experiment 4: Discrimination Learning

In Experiment 1 trials were separated by a 24-hour ITI. Following the daily trial in the alley the animal was returned to the home cage and fed shortly thereafter. Consider the following sequence of events: The animal is nonrewarded in the alley on Trial T, returned to its home cage and fed, and run in the alley on the following day, Trial T + 1. On Trial T + 1 the animal remembers being nonrewarded on Trial T despite the intervening home cage feeding. By the same token, on being returned to the home cage following nonreward on Trial T, the animal remembers being fed in the home cage on the previous day. The example, while crude, is instructive. It suggests that what the animal will remember in a particular situation is determined by his last experience in that situation. *Let us go further and insist that the animal will remember perfectly what occurred previously in a particular situation only if that situation is repeated identically.* According to this rule, a failure to remember a particular situation in whole or in part is seen in terms of a failure of that situation to be repeated in all essentials. This principle is very powerful in the context of the runway situation. By means of it one may explain a number of phenomena as, for example, the fact that following a shift from large reward to small reward behavior is initially disrupted in the goalward portion of the alley. This disruption extends to the start portion on later trials. We are concerned here, however, only with applying the principle to discrimination learning.

Consider an animal rewarded in the presence of a black stimulus and nonrewarded in the presence of a white stimulus. Assume further that black and white are maximally dissimilar so that the animal will not tend to become "confused" when remembering that reward was associated with black and nonreward with white. In this example, then, when placed before black stimuli the animal remembers reward and when placed in front of the white stimulus it remembers nonreward. Employing these notions it is possible to show that it is rather easy for the animal to learn double alternation behavior when reward and nonreward are double alternated. The double alternation situation was chosen because, as will be seen, a situation of that sort not only presents a difficult test for the memory notion, but also seems difficult to explain on the basis of other views of discrimination learning.

The conditions under which double alternation behavior will not appear or appear only poorly (e.g., Bloom & Capaldi, 1961) are those specified by the hypothesis described above. Essentially the external stimuli are held rather constant from trial to trial (e.g., black alley) and reward and nonreward are double alternated, i.e., RRNNRRNN, etc. This is shown schematically below. Let us examine this situation in terms of the tabulated model below:

	Trial							
	1	2	3	4	5	6	7	8
Alley color (B = black)	B	B	B	B	B	B	B	B
Reward outcome (R or N)	R	R	N	N	R	R	N	N

It can be seen above that the memory of reward (S_R) occurs on a rewarded trial such as Trial 2, and on a nonrewarded trial such as Trial 3. Moreover, the memory of nonreward (S_N) occurs on a nonrewarded trial such as Trial 4 and on a rewarded trial such as Trial 5. Thus both S_N and S_R are both rewarded and nonrewarded and double alternation is difficult for the animal.

Let us now examine a situation in which, according to the model described, it should prove relatively easy for the animal to learn to double alternate. In this case, alley color alternates in a simple manner from trial to trial while reward and nonreward double alternate. This is seen in the tabulation below:

	Trial							
	1	2	3	4	5	6	7	8
Alley color (B = black, W = white)	B	W	B	W	B	W	B	W
Reward outcome (R or N)	R	R	N	N	R	R	N	N

First let us satisfy ourselves that if the animal double alternates in this rather unusual situation, it cannot do so on the basis of external stimuli. Note that reward and nonreward occur equally often in black and in white. Consider the situation now from a memory standpoint. What will the animal remember on Trial 3? According to the model described, it will remember S_R from Trial 1 (what last happened in black). Likewise, on Trial 4 it will remember what last happened in white (Trial 2) or S_R. We have seen, then, that on Trials 3 and 4, both of which are nonrewarded, the animal remembers reward or S_R. On Trial 5 the animal remembers S_N from Trial 3. And on Trial 6 the animal remembers nonreward from Trial 4. On rewarded trials, then, the animal remembers nonreward (S_N-R_A). And on nonrewarded trials the animal remembers reward (S_R no R_A). According to the memory model, then, the problem is a rather easy one for the rat. In "Go" "No Go" terms the animal need only learn "when I remember reward, no go" and "when I remember nonreward, go."

A. METHOD

1. Subjects

The Ss, 24 naive male albino rats purchased from Holtzman Co., Madison, Wisconsin, were about 90 days old upon arrival at the laboratory.

2. Apparatus

The apparatus consisted of a black and a white wooden runway, each 43 inches long, 3½ inches wide, and 4 inches deep. Each runway had a start box of the same color, 8 inches long, 2½ inches wide, and 4 inches deep. The start boxes were separated from the runways by a plexiglas guillotine door to form a goal box which contained a small glass furniture coaster for holding 0.045-gm Noyes pellets. Response times were recorded by Hunter Photorelays and Klockounters from the opening of the start box door to a point 12 inches into the runway (start), over the next 15 inches (run), and over the following 12 inches (goal).

In addition to the use of different colors, the alleys and start boxes were distinguished in two ways. The floor of the black start box and alley was smooth, while the floor of the white start box and alley was covered with ½-inch hardware cloth. When the black start box and alley were in use, illumination was provided by a single neon ceiling fixture; when the white start box and alley were in use, the illumination provided by the ceiling fixture was supplemented by that of a translucent acetate sky containing many 7-W light bulbs. The sky was hung 4 feet above the apparatus floor.

3. Procedure

Upon arrival at the laboratory Ss were housed individually and given *ad lib.* access to food and water for 2 days. On the succeeding 12 days Ss were handled and placed on a 12 gm per day maintenance schedule which included twelve 0.045-gm Noyes pellets daily. Two days prior to experimental training each S was individually placed once in each goal box and allowed to eat 12 pellets per placement. One day prior to experimental training each S was given two rewarded (12 pellets) running trials, one in each alley.

Experimental training consisted of a 42-day training phase and a 6-day shift phase (shift phase not reported here). On every day of the training phase each S received four rewarded trials (R) and four nonrewarded trials (N) in a double alternation sequence, RRNNRRNN. In every case the trials were scheduled such that within a 6-day block, half the R trials and half the N trials occurred in each alley. The specific order in which the alleys were used for a given S constituted the variable under investigation.

On each trial S was placed in the start box, and after 2 seconds the start box door was raised. S was allowed 60 seconds to cross an alley section before being forced to do so. On rewarded trials the goal cup contained twelve 0.045-gm Noyes pellets and S was removed after eating. On nonrewarded trials S was confined to the goal box for a time before removal. On the first 25 days of the training phase the nonreward confinement time was 20 seconds; during the remainder of the training phase the nonreward confinement time was 30 seconds.

Ss were taken in squads of eight from the animal colony to the experimental room where they were run in rotations of two. The resulting intertrial interval was 1½–2 minutes. Ss were fed their daily ration 30 minutes to 1 hour after being run.

Recent research suggests that rats may in some way be cued by the reward outcome of the S previously run in the apparatus and that when successive reward outcomes are correlated (i.e., when Ss are run in rotation on the same partial reinforcement schedule) these cues may support pattern running. In the present study, within a squad of eight, Ss were assigned positions in rotations daily such that within all two-day blocks there was no correlation of successive reward outcomes. This statement applies whether the reward outcomes of both alleys are considered or if each alley is considered individually.

During the 42-day acquisition phase there were two experimental conditions. Eighteen Ss were run in one condition (Alt) and six Ss were run in a second condition (Irreg). The two alleys will be referred to as S_1 and S_2. The Alt group received the same alternating order of alleys, S_1 S_2 S_1 S_2 S_1 S_2 S_1 S_2, each day. Group Irreg received six irregular orders of alley presentation. Every order was used once within each block of 6 days. The six orders were S_1 S_1 S_2 S_2 S_2 S_1 S_1 S_2, S_2 S_2 S_1 S_1 S_1 S_2 S_2 S_1, S_2 S_1 S_1 S_2 S_2 S_2 S_1 S_1, S_1 S_2 S_2 S_1 S_1 S_1 S_2 S_2, S_2 S_1 S_1 S_1 S_2 S_2 S_1 S_2, S_1 S_1 S_2 S_1 S_2 S_1 S_2 S_2.

For half the Ss in each group the black alley was used on S_1 trials and the white alley was used on S_2 trials; for the other half the alley assignments were reversed.

B. RESULTS

Since similar results were obtained in each alley section, only total speeds are reported here. Figure 5 shows total running speed for each of the groups over the 42-day training phase on the basis of R trial speeds vs. N trial speeds. As can be seen in Fig. 5, the R and N curves for both groups rise together until Day 13, whereupon the N curves begin to fall rapidly. The N curve for the Alt condition asymptotes at about 1.5 feet/second, about one half the speed on R trials, indicating strong patterning. In the irregular color condition the N curve falls to about 2.6 feet/second indicating weak patterning.

A group (alternating vs. irregular) by trial type (R vs. N) analysis was performed on the data from the last two days of the training phase. Not only

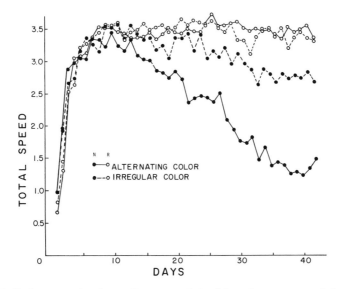

Fig. 5. Performance (total speed) on rewarded trials and on nonrewarded trials for Group Alternating and for Group Irregular on each day of training (Experiment 4).

was the between-groups difference significant $(F = 7.65,\ df = 1/22,\ p < .05)$, but more importantly, the Groups x Trial Type interaction was significant $(F = 23.68,\ df = 1/22,\ p < .005)$. Clearly, then, the alternating group was discriminating R trials from N trials better than the irregular group.

Close inspection of the running speeds obtained on each of the eight trials of acquisition on Days 41 and 42 raises considerable doubt that the discrimination shown in Fig. 5 between R and N trials in the irregular group was in fact discrimination. On the first two trials of the day (which, of course, were R trials) the irregular group ran very rapidly. On the subsequent six trials of the day, however, the irregular group ran less rapidly and nondifferentially. Thus, comparing Trials 1 and 2 (R trials) with Trials 3 and 4 (N trials) in the irregular group yields a significant F $(F = 11.15,\ df = 1/40,\ p < .005)$. The corresponding F for the alternating group was 235.10. However, comparing Trials 5 and 6 (R trials) with Trials 7 and 8 (N trials) in the irregular group yields $F < 1$. In the alternating group the corresponding F was 89.25 $(df = 1/40,\ p < .001)$. Thus, it is possible that in the irregular group a clear discrimination between R and N trials failed to occur. Rather, the Ss in this group may have learned simply to run rapidly on the initial two trials of the day.

C. DISCUSSION

The simple analysis of discrimination learning presented here assumes that the animal will remember what reward (R or N) it received last in the particular alley. On this basis the discrimination shown by the alternating group is understandable. In common sense terms the alternating group need only have learned "if I was rewarded in this alley on the previous trial, I will be nonrewarded on the current trial" and "if I was nonrewarded on the previous trial, I will be rewarded on the current trial." That this sort of mechanism can explain discrimination learning in the more typical discrimination task, i.e., black alley always rewarded, white alley always nonrewarded, is obvious. The presence of discrimination learning in the alternating group in the present investigation, however, is consistent with a quite important generalization within the memory model; namely, the S will remember the reward condition it received in a particular alley when it is returned to that alley.

We think it desirable to indicate one other implication of the present results. The discrimination shown by the alternating group is plainly contrary to the implications of the kind of analysis suggested by Sheffield. According to this view, the sensory consequences of reward and of nonreward persist briefly until the following trial (15 minutes maximum). Since the schedule of reward double alternated, it can be seen that the stimulus aftereffect of reward would occur on both R and N trials, while likewise the stimulus aftereffect of nonreward trials would also occur on R and N trials. Accordingly, from an aftereffects view, no basis for the sort of discrimination shown by the alternating group can exist. What seems to be required, then, is a memory model. A memory model suggests that the stimulus consequence of a particular reward condition will occur on a particular trial only if the external stimuli which accompanied that reward condition are re-presented. Put differently, the internal stimuli which occur on a particular trial depend upon the specific external stimuli which occur on that trial. Expressed in terms of language employed earlier in this chapter, the internal stimuli are external stimulus dependent (memory), not external stimulus independent (aftereffects).

V. Conclusions

According to the classical conception, as represented both in cognitive and stimulus–response theory, expectation is considered to be a slowly increasing function of number of rewarded trials and a slowly decreasing function of number of nonrewarded trials. Insofar as expectation is ultimately represented as an internalized stimulus state of the animal, changes in the internal stimulus environment occur slowly and in small steps over a series of rewarded or nonrewarded trials. In selected instances, regarded as the exception rather than the rule, the classical conception has provided for rapid and extensive changes in the internal stimulus state of the animal, and, particularly of late, has increasingly come to accept this position.

In contrast to the classical model, a memory model has been presented here which described a radically different conceptualization of the behavior and alteration of the animal's internal stimulus environment. Rapid and extensive change in the animal's internal stimulus state as a function of changed conditions of reward was seen to be the rule and not the exception. In connection with this analysis, data have been presented here which simultaneously supported a memory conception and disconfirmed the gradual change, learning conception. Moreover, it has been shown that these data are not the products of experiments which are isolated and discrete in the sense of failing to make contact with previously reported experimental results. On the contrary, much of the data presented here was shown to be continuous with extensive bodies of literature, some of it extending back to the earliest critical days in which phenomena in the area of varied reward came under investigation. Gradual change hypotheses, in their attempt to deal with these data, e.g., percentage-schedule data, either have no explanation for them (e.g., Amsel, 1958), or have explanations which are grossly inadequate (e.g., Lawrence & Festinger, 1962). Further, such hypotheses deal with these data by fractionating and dividing them into various classes, e.g., in terms of artifactual vs. nonartifactual results, as lying within or beyond the scope of the theory, and so on. The sequential hypothesis dealt with these data quite differently. It viewed the data in these areas as continuous and interrelated, and dealt with all of these data simultaneously, employing a single set of coherent principles.

Differences between the memory model and gradual internal stimulus change models cannot be understood apart from the different theoretical strategies each has adopted until now. This point cannot be overemphasized. In the beginning both types of models floundered around, as it were, until some sort of preliminary explanation of at least some classes of varied reward data was evolved. Beyond this point, the theoretical strategies of the two kinds of models differed radically. Frustration perhaps provides the best contrast to the sequential model because it has generated more research and attracted more adherents than various other gradual change models, e.g., dissonance and competing responses. Once the frustration model evolved an explanation of why partial reward occasions greater resistance to extinction than consistent reward, its attention became directed to frustration itself. It was asked, in the main, what other sorts of phenomena can be understood in terms of frustration, e.g., discrimination learning (e.g., MacKinnon & Amsel, 1964). The sequential model, as has been indicated, adopted a different strategy. After having evolved an explanation of the PRE, the sequential model asked not what implications do sequential principles have for other kinds of learning situations, but rather, what implications do other varieties of partial reward situations have for sequential principles.

The difference between the two sorts of strategies is quite concrete and not at all subtle. The PRE has had a tremendous impact on theorizing. Generally speaking, any model which explains the PRE will employ, relative to classical theories (e.g., Hull, 1943), new and different principles. One can assume, as in our view most gradual change models in fact did assume, that the new principles, developed on the basis of the PRE alone, were sufficiently general so as to safely allow application elsewhere, in discrimination learning, double alleys, and so on. The implicit assumption here seems to have been that one could all but ignore other conditions of varied reinforcement, percentage, schedule, and so on, without running the risk of premature generalization. A diametrically opposite assumption was entertained within sequential theory. The judgment was that the PRE itself was compatible with a variety of theoretical conclusions—witness the great procession of hypotheses which have explained the PRE. Morever, even within the context of a particular explanation of the PRE, a variety of alternative conclusions was possible. In the face of these facts, it was assumed within sequential theory that other conditions of partial reward necessarily had to be investigated before one could identify the specific form our general principles should take. To put matters bluntly, the PRE itself suggested to all of us, gradual change theorists and sequential theorists alike, that we had to theorize in terms of internal stimuli. That, however, was only the first step. Percentage-schedule data, small-trial PRE data, and the like taught us what particular assumptions had to be made in connection with these internal stimuli. In our view, it was this second step that was skipped by gradual change models. Either because they were anxious to return as much as possible to traditional concepts (for instance, r_g) or to traditional learning situations (for example, discrimination learning), or because interest began to focus primarily upon the processes evolved from an analysis of the PRE, i.e., frustration, various gradual change models ignored what other varieties of partial reinforcement situations had to offer. In our view, this constituted a faulty strategy; what we have called the second step has proved to be at least as important as the first step. It is because gradual change models skipped the second step that they now find themselves in the position of postulating distinctions of various sorts, e.g., between regular and irregular schedules, or postulating various two-process accounts of the PRE, or radically modifying previous assumptions (e.g., Amsel *et al.,* 1968).

Only recently has the sequential model attempted to explain other sorts of learning effects, e.g., negative contrast effects (e.g., Capaldi & Lynch, 1967; Capaldi & Ziff, 1971), reward magnitude effects (e.g., Capaldi & Minkoff, 1971; Leonard, 1969), successive acquisitions and extinctions (Capaldi *et al.,* 1968b), and so on (see Capaldi, 1967). Experiment 4 reported here represents an initial attempt to apply sequential principles to discrimination learning. Actually, we are still not fully confident that enough is known about how memories operate to justify turning to these more complex, or at any rate different cases, particularly discrimination learning. In any event, in the last 3 years or so sequential theory has finally embarked on this task and only the future can reveal what sorts of difficulties may be in store.

ACKNOWLEDGMENTS

Considerable appreciation is expressed to those who aided the completion of this work in its various stages. Most particularly we would like to thank Elizabeth D. Capaldi, Robert C. Godbout and David R. Ziff.

REFERENCES

Amsel, A. The role of frustrative nonreward in noncontinuous reward situations. *Psychological Bulletin,* 1958, 55, 102–119.

Amsel, A. Partial reinforcement effects on vigor and persistence. In K. W. Spence & J. T. Spence (Eds.), *The psychology of learning and motivation.* Vol. 1. New York: Academic Press, 1967. Pp. 1–65.

Amsel, A., Hug, J. J., & Surridge, C. T. Number of food pellets, goal approaches, and the partial reinforcement effect after minimal acquisition. *Journal of Experimental Psychology,* 1968, 77, 530–534.

Amsel, A., & Ward, J. S. Frustration and persistence: Resistance to discrimination following prior experience with the discriminanda. *Psychological Monographs,* 1965, 79 (4, Whole No. 597).

Anderson, N. Comparison of different populations: Resistance to extinction and transfer. *Psychological Review*, 1963, 70, 162–179.

Bacon, W. E. Partial reinforcement extinction effect following different amounts of training. *Journal of Comparative and Physiological Psychology*, 1962, 55, 998–1003.

Black, R. W., & Spence, K. W. Effects of intertrial reinforcement on resistance to extinction following extended training. *Journal of Experimental Psychology*, 1965, 70, 559–563.

Bloom, J. M., & Capaldi, E. J. The behavior of rats in relation to complex patterns of partial reinforcement. *Journal of Comparative and Physiological Psychology*, 1961, 54, 261–265.

Bloom, J. M., & Malone, P. Single alternation patterning without a trace for blame. *Psychonomic Science*, 1968, 11, 535–336.

Capaldi, E. J. Effect of N-length, number of different N-lengths and number of reinforcements on resistance to extinction. *Journal of Experimental Psychology*, 1964, 68, 230–239.

Capaldi, E. J. Partial reinforcement: A hypothesis of sequential effects. *Psychological Review*, 1966, 73, 459–477.

Capaldi, E. J. A sequential hypothesis of instrumental learning. In K. W. Spence & J. T. Spence (Eds.), *The psychology of learning and motivation*. Vol. 1. New York: Academic Press, 1967. Pp. 67–156.

Capaldi, E. J., & Hart, D. Influence of a small number of partial reinforced training trials on resistance to extinction. *Journal of Experimental Psychology*, 1962, 64, 166–171.

Capaldi, E. J., Lanier, A. T., & Godbout, R. C. Reward schedule effects following severely limited acquisition training. *Journal of Experimental Psychology*, 1968, 78, 521–524. (a)

Capaldi, E. J., Leonard, D. W., & Ksir, C. A reexamination of extinction rate in successive acquisitions and extinctions. *Journal of Comparative and Physiological Psychology*, 1968, 66, 128–132. (b)

Capaldi, E. J., & Lynch, D. Repeated shifts in reward magnitude: Evidence in favor of an associational and absolute (non-contextual) interpretation. *Journal of Experimental Psychology*, 1967, 75, 226–235.

Capaldi, E. J., & Lynch, A. D. Magnitude of partial reward and resistance to extinction: Effect of N-R transitions. *Journal of Comparative and Physiological Psychology*, 1968, 65, 179–181.

Capaldi, E. J., & Minkoff, R. Change in the stimulus produced by nonreward as a function of time. *Psychonomic Science*, 1966, 6, 321–322.

Capaldi, E. J., & Minkoff, R. Reward schedule effects at a relatively long intertrial interval. *Psychonomic Science*, 1967, 9, 169–170.

Capaldi, E. J., & Minkoff, R. Influence of order of occurrence of nonreward and large and small reward on acquisition and extinction. *Journal of Experimental Psychology*, 1971, in press.

Capaldi, E. J., & Stanley, L. R. Percentage of reward vs. N-length in the runway. *Psychonomic Science*, 1965, 3, 263–264.

Capaldi, E. J., & Wargo, P. Effect of transitions from non-reinforced to reinforced trials under spaced trial conditions. *Journal of Experimental Psychology*, 1963, 65, 318–319.

Capaldi, E. J., & Ziff, D. R. Effect of schedule of partial reward on the negative contrast effect. *Journal of Comparative and Physiological Psychology*, 1971, in press.

Crespi, L. P. Amount of reinforcement and level of performance. *Psychological Review*, 1944, 51, 341–357.

Godbout, R. C., Ziff, D. R., & Capaldi, E. J. Effect of several reward exposure procedures on the small trial PRE. *Psychonomic Science,* 1968, **13**, 153-154.

Gonzalez, R. C., Bainbridge, P., & Bitterman, M. E. Discrete trials lever pressing in the rat as a function of pattern of reinforcement, effortfulness of response, and amount of reward. *Journal of Comparative and Physiological Psychology,* 1966, **61**, 110-122.

Gonzalez, R. C., & Bitterman, M. E. Resistance to extinction in the rat as a function of percentage and distribution of reinforcement. *Journal of Comparative and Physiological Psychology,* 1964, **58**, 258-263.

Gonzalez, R. C., & Bitterman, M. E. Spaced-trials partial reinforcement effect as a function of contrast. *Journal of Comparative and Physiological Psychology,* 1969, **67**, 94-103.

Grosslight, J. H., Hall, J. R., & Murnin, J. Patterning effects in partial reinforcement. *Journal of Experimental Psychology,* 1953, **46**, 103-106.

Grosslight, J. H., & Radlow, R. Patterning effect of the nonreinforcement-reinforcement sequence involving a single nonreinforced trial. *Journal of Comparative and Physiological Psychology,* 1957, **50**, 23-25.

Hanford, P. V., & Zimmerman, J. Differential running in rats under an alternating (FR 2) schedule in an automated runway. *Psychonomic Science,* 1969, **14**, 107-108.

Harris, J. H., & Thomas, G. J. Learning single alternation of running speeds in a runway without handling between trials. *Psychonomic Science,* 1966, **6**, 329-330.

Howlett, J. C., & Sheldon, M. H. Effects of partial delay of reinforcement following a small number of acquisition trials. *Psychonomic Science,* 1968, **11**, 259-260.

Hull, C. L. Goal attraction and directing ideas conceived as habit phenomena. *Psychological Review,* 1931, **38**, 487-506.

Hull, C. L. *Principles of Behavior.* New York: Appleton, 1943.

Hull, C. L. *A behavior system.* New Haven, Conn.: Yale Univ. Press, 1952.

Hulse, S. H. Amount and percentage of reinforcement and duration of goal confinement in conditioning and extinction. *Journal of Experimental Psychology,* 1958, **56**, 48-57.

Lawrence, D. H., & Festinger, L. *Deterrents and reinforcement.* Stanford, Calif.: Stanford Univ. Press, 1962.

Leonard, D. W. Amount and sequence of reward in partial and continuous reinforcement. *Journal of Comparative and Physiological Psychology,* 1969, **67**, 204-211.

Logan, F. A. Incentive theory and changes in reward. In K. W. Spence and J. T. Spence (Eds.), *The psychology of learning and motivation.* Vol. 2. New York: Academic Press, 1968. Pp. 1-30.

McCain, G. Partial reinforcement effects following a small number of acquisition trials. *Psychonomic Monograph Supplement,* 1966, **1**, 251-270.

McCain, G. The partial reinforcement effect after minimal acquisition: Single pellet reward, spaced trials. *Psychonomic Science,* 1969, **15**, 146.

MacKinnon, J. R., & Amsel, A. Magnitude of the frustration effect as a function of confinement and detention in the frustrating situation. *Journal of Experimental Psychology,* 1964, **67**, 468-474.

Padilla, A. M. A few acquisition trials: Effects of magnitude and percent reward. *Psychonomic Science,* 1967, **9**, 241-242.

Robbins, D., Chait, H., & Weinstock, S. Effects of nonreinforcement on running behavior during acquisition, extinction, and re-acquisition. *Journal of Comparative and Physiological Psychology,* 1968, **66**, 699-706.

Sheffield, V. F. Extinction as a function of partial reinforcement and distribution of practice. *Journal of Experimental Psychology,* 1949, **39**, 511-526.

Spear, N. E., Hill, W. F., & O'Sullivan, D. J. Acquisition and extinction after initial trials without reward. *Journal of Experimental Psychology,* 1965, **69**, 25-29.

Spear, N. E., & Spitzner, J. H. Effect of initial nonrewarded trials: Factors responsible for increased resistance to extinction. *Journal of Experimental Psychology,* 1967, **74,** 525–537.

Spence, K. W. *Behavior theory and conditioning.* New Haven, Conn.: Yale Univ. Press, 1956.

Spence, K. W. *Behavior theory and learning.* Englewood Cliffs, N.J.: Prentice-Hall, 1960.

Spence, K. W., Platt, J. R., & Matsumoto, R. Intertrial reinforcement and the partial reinforcement effect as a function of number of training trials. *Psychonomic Science,* 1965, **3,** 205–206.

Spivey, J. E. Resistance to extinction as a function of number of N-R transitions and percentage of reinforcement. *Journal of Experimental Psychology,* 1967, **75,** 43–48.

Surridge, C. T., & Amsel, A. Performance under a single alternation schedule of reinforcement at 24-hour intertrial interval. *Psychonomic Science,* 1965, **3,** 131–132.

Surridge, C. T., & Amsel, A. Acquisition and extinction under single alternation and random partial reinforcement conditions with 24-hour intertrial interval. *Journal of Experimental Psychology,* 1966, **72,** 361–368.

Surridge, C. T., Rashotte, M. E., & Amsel, A. Resistance to extinction of a running response after a small number of partially rewarded trials. *Psychonomic Science,* 1967, **7,** 31–32.

Uhl, C. N., & Young, A. G. Resistance to extinction as a function of incentive percentage of reinforcement, and number of nonreinforced trials. *Journal of Experimental Psychology,* 1967, **73,** 556–564.

Wagner, A. R. Effects of amount and percentage of reinforcement and number of acquisition trials on conditioning and extinction. *Journal of Experimental Psychology,* 1961, **62,** 234–242.

Weinstock, S. Resistance to extinction of a running response following partial reinforcement under widely spaced trials. *Journal of Comparative and Physiological Psychology,* 1954, **47,** 318–322.

Weinstock, S. Acquisition and extinction of a partially reinforced running response at a 24-hour intertrial interval. *Journal of Experimental Psychology,* 1958, **56,** 151–158.

Chapter 4

THE ROLE OF INTERFERENCE
IN ASSOCIATION OVER A DELAY[1]

Sam Revusky

I. Introduction

This chapter is concerned with the theoretical implications of recent findings that animals can learn even when two elements of a single learning paradigm are separated by a substantial temporal interval. One example of such learning over a long delay is the Garcia Effect (Garcia, Kimeldorf, & Koelling, 1955) in which the discriminative stimulus and the response are separated from a punishment by a number of hours. An animal is made to consume a flavored substance and the punishment is toxicosis induced as much as 12 hours later by injection of a toxin or by X-irradiation. When the preference for the flavored substance is tested, the animal exhibits an aversion to the flavor even after it has recovered from the toxicosis. Furthermore, such learning is not at all difficult to obtain with but a single pairing between the flavored substance and the toxicosis. It is presupposed in this chapter that the Garcia Effect is to be attributed to a direct association between the flavor and the toxicosis, for no mediational processes can reasonably be invoked to bridge the temporal gap unless memories are defined as mediational processes. The numerous experiments involving the Garcia Effect which justify these presuppositions are summarized in a review by Revusky and Garcia (1970).

[1] Because this chapter is really a variation on a theme by John Garcia, it is dedicated to him. Bow Tong Revusky skillfully edited and rewrote many sections. One of the theoretical improvements for which she is responsible is the derivation of the proximity corollary from the concurrent interference principle in Section III, F; originally, proximity was treated as a separate principle. Werner Honig improved the chapter a great deal, and I am grateful to him for bringing his expertise in stimulus control to bear in his critique of an earlier version. In a sense, this chapter is a development of the theory of stimulus control to cover associations over long delays. Rubin Gotesky and Henry James also made suggestions which I incorporated into this chapter. The writing of this chapter and the experimentation reported in Section VI were supported by United States Public Health Service Grants MH 16423 and MH 16643.

The Capaldi Effect is another case in which learning occurs although two elements of the learning paradigm are separated by a long delay (Capaldi, 1967). Rats, rewarded on alternate trials in a runway, learn to run more slowly on unrewarded trials than on rewarded trials even with a 24-hour intertrial interval. In Chapter 3, Capaldi has convincingly interpreted this result as evidence that the reward outcome of the preceding trial may be a discriminative stimulus for the following trial. Recently, Martin Pschirrer (unpublished doctoral work in the author's laboratory) has supplied additional evidence in favor of this interpretation. In his first experiment, one group of rats was subjected to a sequence of three runway trials repeated indefinitely; the first was rewarded with milk, the second was rewarded with chow pellets, and the third was not rewarded. For another group, the order of milk and pellets was interchanged. In both groups, the rats learned to run slowest on the unrewarded trial despite an intertrial interval of over 15 minutes. This result showed that the particular type of reward can be a discriminative stimulus for the reward outcome of the following trial. The remote possibility that the rats were learning to count two successive rewards without utilizing the type of reward substance as a cue was precluded by a second experiment. Pschirrer made the correct goal box of a T-maze depend on the type of reward received on the preceding trial. Using a modified correction procedure, six rats were subjected to a quasi-random sequence of milk and pellet rewards which were equally likely to be dispensed in either goal box. These reward outcomes were discriminative stimuli for the following trial; the left goal box was correct if pellets had been the preceding reward and the right goal box was correct if milk had been the preceding reward (or vice versa). After about 700 trials with an intertrial interval of over 3 minutes, the rats were correct on over 80% of the trials, where 50% is the chance level. This is evidence that the outcome of the preceding trial can function not only to determine running speed, as had been demonstrated by the runway experiment, but also as a discriminative stimulus for the selection of one of two responses. Earlier work by Petrinovich and Bolles (1957) and by Petrinovich, Bradford, and McGaugh (1965) showed that rats can use the type of response emitted on the preceding trial as a discriminative stimulus; that is, rats can learn to alternate in a T-maze even when the intertrial interval is a number of hours.

The above findings blatantly contradict the notion that two events must occur nearly in temporal contiguity if an animal is to associate between them. Since this notion has been tacitly accepted in nearly all of the traditional learning theories, some reconceptualization is necessary. The gist of the reconceptualization proposed here is that temporal contiguity is not at all necessary for association to occur. Presumably, the reason associations over long delays usually do not occur is that the events which are bound to occur during the delay produce interference which prevents the desired delayed association from occurring. In the cases in which long delay learning does occur, this interference is somehow circumvented. Below, this approach will be developed in detail and will be shown to apply both to conventional learning situations and to long delay learning situations.

II. Associative Memory and Retentive Memory

An animal that learns in spite of a delay between two elements of a single learning paradigm will be considered in this chapter to have exhibited "associative memory" over that delay. "Associative memory" is a term which once was frequently used by psychologists, but has since fallen into disuse. The term is used here because it is a convenient label to designate all instances of association over a delay regardless of the different learning paradigms and experimental procedures which may be involved. Furthermore, it permits a distinction between the topic of concern here and the type of memory which has usually been studied by animal psychologists. The latter type of memory may be called retentive memory. It is investigated by training an animal, then preventing it from practicing what it has learned for some interval of time, and then testing it for retention of the effects of the original training. Retentive memory over the period between training and testing is considered to have been exhibited if the animal still shows the effects of training. In other words, retentive memory is concerned with the retention of a learned behavior while associative memory is concerned with learning itself. For this reason, before dealing with associative memory, we must first discuss learning.

Learning may be considered the control of behavior by the previous sequential occurrence of two events, where an event is anything to which an animal can respond (including its own behavior). The first of these events will be called E-pre and the second will be called E-post. In classical conditioning, E-pre is a CS and E-post is a UCS; if the reaction to the CS changes because it has been followed by the UCS, classical conditioning is said to have occurred. In instrumental conditioning, E-pre is a response and E-post is a reward or punishment; if the rate of response changes because the response has been followed by a reward or punishment, instrumental conditioning is said to have occurred. The case of discrimination learning is a bit difficult to place in this rubric because three elements are involved in the paradigm: the discriminative stimulus, the response, and the reinforcement (or punishment). Discrimination learning may be conceptualized either as an association between the discriminative stimulus (E-pre) and a compound (E-post) of the response and reinforcement or, alternatively, as an association between a compound (E-pre) of the discriminative stimulus and the response with the reinforcement (E-post). Usually, however, it is convenient and adequate to ignore the response and to treat discrimination learning as the association of a discriminative stimulus (E-pre) with a reinforcement (E-post). If the probability of a response following a discriminative stimulus changes as a function of the reinforcement contingency with which the stimulus is correlated, discrimination learning is said to have occurred.

A description of learning in terms of an association between E-pre and E-post permits a more precise distinction between associative memory and retentive memory. Associative memory is concerned with the effects of a time interval between E-pre and E-post on learning. In common sense language, an animal must remember E-pre at the time E-post occurs if the two events are to become associated. Retentive memory concerns the effect of the time interval between training and testing. In common sense language, an animal must remember an association it has already learned. Note that the distinction between associative memory and retentive memory does not imply that retentive memory is not involved in experiments primarily concerned with associative memory. Indeed, there must always be a time interval between training and testing even if it is not a matter of primary experimental concern. Nor can the distinction imply a denial that association occurs in experiments about retentive memory, for association occurs in all learning.

A. The Traditional Approach

The traditional position has been that associative memory is extremely short lasting, while retentive memory is extremely long lasting. Compare the treatments of the two types of memory by one of the most highly regarded textbooks on learning of the last decade, Kimble's revision (1961) of *Conditioning and Learning* by Hilgard and Marquis. This is what Kimble states about retentive memory:

> The numerous demonstrations of the great resistance of CR's to forgetting include reports of the retention of conditioned motor responses by sheep for 2 years; conditioned eyelid reactions in dogs for 16 months; conditioned eyelid reactions in man for 20 weeks and for 19 months; conditioned flexion reactions in the dog for 30 months; conditioned salivation in man for 16 weeks; various responses in dogs for 6 months and a pecking response in pigeons for 4 years (Kimble, 1961, p. 281; Kimble's citations of particular papers have been omitted).

Now compare the above with what the same text states about associative memory:

> Experiments designed to determine the exact form of the delay-of-reinforcement gradient have led to successive revisions downward of the estimated amount of time by which reinforcement can be delayed if learning is to occur. At the present time it seems unlikely that learning can take place at all with delays of more than a few seconds. This statement applies to negative as well as positive reinforcers. Instances of learning with protracted delays of reward are always cases where immediate secondary reinforcement occurs. . . .

> The interstimulus interval refers to the time between onset of a conditioned

stimulus and the onset of an unconditioned stimulus. Studies in which the interstimulus interval has been manipulated have shown that the optimal temporal separation of CS and UCS is about .5 seconds, with the CS appearing first. Either longer or shorter intervals lead to poorer learning. . . .

The optimal value, .5 seconds, seems to apply as well to long latency, autonomically controlled responses, such as the GSR, as it does to short latency skeletal reactions such as the eyeblink . . . (Kimble, 1961, pp. 165–166).

From Kimble's very accurate summary of the information available a decade ago, one might expect retentive memory and associative memory to be entirely different, perhaps even to depend on different nervous systems. Learning theories based upon this information reflected this apparent difference between the two types of memory. Indeed, there has been a tendency for theorists to exaggerate this difference. Some theorists concerned with retentive memory have explained away all failures in retentive memory as due to extraneous factors, such as proactive and retroactive interference. At the same time, theorists concerned with associative memory have explained away the few demonstrations of associations over delays longer than a few seconds in terms of extraneous factors, such as secondary reinforcement or mediating chains of behavior. This was the 1947 position of Spence, which has been eloquently summarized by Deese and Hulse (1967): "Spence notes, however, that if it were possible to remove all conditioned reinforcers during the delay period, the gradient of reinforcement ought almost to vanish!" In terms of the nomenclature used here, Spence supposed that, in principle, associative memory is practically nonexistent; in this he expressed the dominant view among American learning theorists. These beliefs are so strongly entrenched that it is small wonder that terms which carry an aura of transience, such as "stimulus trace" and "short-term memory," have become more widely used than the more neutral term "associative memory."

B. The Proposed Approach

The proposed approach is almost the exact opposite of the traditional approach: *Associative memory is hypothesized to be in principle long lasting.* It is not denied here that in practice associative memory is short lasting when the usual experimental procedures of animal psychology are used. However, these usual results are attributed not to a decay in the memory process itself but to associative interference. That is, the events of the reference association (the association designated for study by the experimenter) become associated with extraneous events occurring during a delay; and it is hypothesized that such associations interfere with the formation of the reference association. All delays are assumed to contain potentially interfering events because experimental control is far from perfect in psychology; even if all exteroceptive stimuli during an interval were to be successfully removed, an animal would be likely to emit movements and thus produce interfering events. An increase in the delay is bound to produce an increase in the number of these potentially interfering events. Thus when increase of a delay interval prevents an association between E-pre and E-post it is hypothesized that either E-pre or E-post or both have become associated with intervening events. Presumably, if intervening events did not occur, association over indefinitely long delays would occur as readily as association over extremely short delays. Later on, it will be argued that when long delay learning occurs this interference process is circumvented because the E-pre and the E-post belong to a class of events that are less likely to become associated with the intervening events than with each other.

In other words, the proposed approach suggests that the animal always learns something, but not necessarily what the experimenter wants it to learn. The experimenter's concern is with an arbitrary reference association between a particular E-pre and a particular E-post. The experiment can have one of two outcomes: (1) the reference association between E-pre and E-post can be learned; or (2) the reference association can fail to be learned because E-pre and/or E-post separately become associated with other events and these interfering associations are numerous or potent enough to prevent the reference association from being learned. Given this point of view, the study of associative memory becomes part of the study of associative learning; the problem is to determine which of a number of possible associations will occur and how they will interfere with each other. Note that the only specific assumption essential for implementation of this approach is that associative interference occurs.

The terms "E-pre" and "E-post" draw attention away from the differences between such different types of learning as classical and instrumental conditioning. Although these differences are not denied, they are considered irrelevant in the present context. It is assumed that the associative process is the same whether the learning paradigm involves association of a CS with a delayed UCS or the association of an instrumental response with a delayed reward or punishment. In order to deal with this single process at the maximum level of generality, a single vocabulary must be used for all types of learning. This strategy results in a theory that is deliberately incomplete; it deals only with the animal's formation of associations and not with the nature of the animal's reactions to the associations. As in Pavlovian theory, the main role of behavior is as an indicator that an association has been formed. This leaves the present approach open to an adaptation of Guthrie's objection (1952) to Tolman's theory: It leaves the rat buried in associations because it has no mechanism by which to translate associations into activity. The defense against this objection is that such problems, which the proposed approach cannot solve and which will only produce confusion, are explicitly eliminated from its scope. After all, no theory should be expected to explain everything. I hope it will become apparent that this emphasis on the similarities between different types of learning leads to generality which could not otherwise be obtained.

Before a detailed exposition of the proposed approach, two intuitive arguments in its favor will be offered.

1. The Ability to Store Information

Common to all memory processes is the storage of information over an interval of time. The traditional presupposition, as implied by the term "stimulus trace," has been that this storage is very inefficient in the case of associative memory. In the case of retentive memory, storage has been supposed to be very efficient. Extrapolating from what is known about the effect of complexity upon retentive memory, a case can be made that the traditional view is no more reasonable than its opposite: that storage should be longer lasting for associative memory than for retentive memory. Associative memory involves the storage of less complex information, a single E-pre, than retentive memory which involves the storage of an E-pre, an E-post, and the relationship between them. Thus it seems unlikely that retentive memory is more long lasting than associative memory in practice because of any limitation in the storage process itself; factors which I define as extraneous must be involved. The extraneous factor to be emphasized here is interference, but it is cautioned that the differences between typical associative memory procedures and typical retentive memory procedures are great enough that the involvement of still other factors cannot be precluded; however, these other possible factors are not of concern here.

2. A Parallel with Physics

It seems intuitively compelling to most psychologists that temporal contiguity is necessary if associations are to occur. Perhaps many readers will be more willing to discard this intuitive notion if they are reminded that physicists have found it profitable to discard a still more compelling intuitive notion. On the basis of daily experience, we know that to move one object with another, either the objects must touch or additional objects must bridge the spatial gap. To a common sense shaped by such experience, it is outrageous to suppose that force is transmitted directly from one object to another through a vacuum. Yet this supposition was central to Newton's theory of gravitation. To appease common sense, some physicists postulated an "ether" inside all vacuums to permit the transmission of force (as well as the propagation of light waves). With additional experience, physicists became willing to suppose that force can travel directly through a vacuum. The theory that animals can associate directly over a temporal gap does not require nearly as great a leap of faith as transmission of force through a vacuum because there is an obvious mechanism for such association: the nervous system. Since the nervous system is capable of remembering a past performance for indefinitely long periods, there is no reason not to expect it to remember an E-pre. Indeed, the same students of animal learning, who suppose that the associative memory of an animal is essentially nonexistent, use electronic or electromagnetic switching devices with indefinitely long associative memories to administer their experimental procedures. It is unreasonable to suppose that the far more numerous and far more complex switching circuits of the nervous system cannot do the same.

C. PROPERTIES OF PROPOSED APPROACH

Perhaps the novelty of the suggested approach to associative memory is being exaggerated here. It actually is reminiscent of the interference approaches to forgetting. Of course, forgetting is the obverse of retentive memory. Both the proposed approach and interference theories of forgetting presuppose that memory is extremely long lasting and that the detrimental effects of an increased time delay are due to increases in the strength and number of interfering associations. Nevertheless, important differences between the two approaches emerge because one is concerned with associative memory and the other is concerned with retentive memory. To deal with associative memory, the primary concern must be with interference which develops during training; the effects of an increase in the time interval between E-pre and E-post are explained in terms of an increase in the number of potentially interfering events produced by lengthening the interval. This type of interference will be called concurrent interference to distinguish it from the types of interference emphasized in theories about retentive memory. The latter types of interference are proactive interference, which develops prior to training, and retroactive interference, which develops after training. The distinction between the present approach to associative memory and interference approaches to retentive memory will tend to blur at times. The present exposition will occasionally tacitly refer to proactive interference as a determinant of the magnitude of concurrent interference. (Obviously, retroactive interference is bound to be irrelevant to associative memory because it occurs after training by definition.) Conceivably, interference theories of retentive memory may find occasion to invoke concurrent interference. Perhaps eventually both associative memory and retentive memory will be explained by a single, very general theory. But at the present time, it seems more profitable to emphasize the distinction between them.

III. Concurrent Interference

A. DEFINITIONS

Although the associations with which this theory is concerned are in the forward direction in time, there is no assumption that E-pre and E-post are different types of events. That E-pre must precede E-post is a definition and not an assumption. That is, when the concern is with an association between Event E and some later event, Event E is called E-pre; when the concern is with the association of some earlier event with Event E, Event E is called E-post. This is no different from the practice in instrumental learning theory of calling the same event either a discriminative stimulus or a secondary reinforcer depending on whether the concern is with its relationship to the behavior it precedes or to the behavior it follows.

Of course, there may be practical experimental constraints on the type of event which may be selected as E-pre or as E-post. For instance, it usually makes little sense to utilize painful electrical shock as E-pre and a light as E-post in a classical conditioning experiment (although it is not entirely inconceivable that such a procedure might produce an anticipatory visual reaction to the shock, particularly if the light were very bright). However, although certain selections of E-pre and E-post may be impractical for a variety of reasons, these reasons are outside the scope of the proposed theory. It must be admitted, however, that at a nearly metaphysical level, there is a potential problem because the theory is about associations and it is quite conceivable that associations may occur under conditions in which they cannot be demonstrated by means of known behavioral techniques. This was the case for stimulus-stimulus associations before the sensory preconditioning technique was developed. But this type of problem will not have any practical ramifications in the present chapter. When experiments are described which are relevant to the analysis of associative memory, the selection of events as E-pre or E-post will meet the usual experimental criteria of reasonableness.

In order to describe the principle of concurrent interference, a distinction must be made between the events of the reference association, E-pre and E-post, and the events which may produce interference. These potentially interfering events will be called E-pre-X and E-post-X. Essentially, E-pre-X is an event that is not part of the reference association and occurs prior to E-post; thus it is possible for E-pre-X and E-post to become associated. Similarly, E-post-X is some event other than E-post that occurs after E-pre. Note that these definitions do not require that E-pre-X's and E-post-X's occur during the interval between E-pre and E-post. Conceivably, an E-pre-X may precede E-pre and an E-post-X may follow E-post.

Training is defined as the presentation of E-post after E-pre occurs in order to produce an association between them. Both E-pre-X and E-post-X are defined specifically as events which must occur during training, and interference which develops prior to training (proactive interference) is eliminated from the scope of the proposed theory. Of course, this not a denial that proactive interference exists.

B. The Concurrent Interference Principle Itself

This is the proposed concurrent interference principle: *The probability and/or strength of an association between any E-pre and E-post decreases as a function of the strength and number of the following types of associations: (1) associations of E-pre-X with E-post and (2) associations of E-pre with E-post-X.*

Note that since there is no inherent difference between the reference association and an interfering association, this principle also implies that if a reference association becomes very strong, it may decrease the strength of interfering associations.

1. All-or-None Interference Is Not Presupposed

This formulation of the concurrent interference principle does not imply that if interference occurs, it must be complete. Interfering associations are presumed to weaken the reference association, but in many situations it may require a great deal of interference to eliminate the reference association entirely. In other words, it is not denied that a number of E-pre's and E-pre-X's can become associated with a single E-post or that a single E-pre can become associated with a number of E-post's and E-post-X's. However, there is a sort of mental law of diminishing returns: The more associations in which a single event becomes involved, the weaker each of the associations will tend to be.

2. Exclusion of Secondary Reinforcement

For the time being, I would like to omit from the scope of the concurrent interference principle the special case in which both E-pre and E-post become associated with a single event which occurs during the interval between them. In this excluded case, the intervening event takes on the role of an E-post-X relative to E-pre and of an E-pre-X relative to E-post. Such an arrangement usually facilitates learning over a delay. For instance, in the case of instrumental learning, where E-pre is a response and E-post is a primary reinforcer, the intervening event may be a neutral exteroceptive stimulus. Many findings show that if the neutral stimulus becomes established as a secondary reinforcer by virtue of its association with the primary reinforcer (E-post), it will increase the probability of the response (E-pre) which it follows. If the concurrent interference principle were left unqualified, it would imply the opposite outcome because the neutral stimulus enters into two interfering associations which are bound to weaken the reference association. Later in this chapter (Section IV, B) an attempt will be made to show that this apparent contradiction is not really damaging to the proposed theory.

C. Experimental Evidence for Concurrent Interference

The principle of concurrent interference has often been a bridesmaid, but only now, in this chapter, has it become a bride. That is, it has been used as a secondary explanatory principle in one form or another by a variety of investigators, but it is only here that it is considered to be a general principle and central to a theory. In the following sections evidence in favor of concurrent interference will be gleaned from a variety of sources, but no attempt will be made to elucidate the original theoretical contexts in which this evidence was presented. Thus whenever any reference is made to an investigation which seems to confirm the concurrent interference principle, there is no implication that the principle was accepted by the investigator in the form in which it is espoused here.

All the evidence to be supplied will be based on the first type of concurrent interference. For analytical purposes consider the simplified case in which two events, E-pre and E-pre-X, occur prior to E-post. For the concurrent interference principle to be confirmed, there must be an inverse relationship between the strength of the interfering association between E-pre-X and E-post and the strength of the reference association between E-pre and E-post. The evidence for this inverse relationship will be considered in terms of three factors which affect associative strength. These are: (1) the psychophysical characteristics of E-pre and E pre-X, (2) the animal's prior history with E-pre and E-pre-X, and (3) the relevance of E-pre and of E-pre-X to E-post. The third factor, "relevance," is a relatively new concept which has an important secondary role in the proposed theory and will be explained later. Each of these three factors will be discussed in terms of its implications for concurrent interference, and some supporting evidence will be summarized.

1. Psychophysical Characteristics of E-pre and E-pre-X

Various psychophysical characteristics affect the readiness with which an E-pre-X will become associated with an E-post. The concurrent interference principle implies that the readiness with which such associations are formed is the same as the readiness with which interference is produced.

a. Examples from Classical Conditioning. Pavlov (1927) defined a strong CS as one which readily becomes associated with a UCS; it begins to elicit a CR rapidly and eventually elicits a larger CR as compared with a weak CS. The strength of a CS was found to be an increasing function of CS intensity and also tended to be a function of CS modality. Obviously, the strength of a CS according to Pavlov corresponds to the associative strength of an E-pre or an E-pre-X in the present nomenclature.

Pavlov's phenomenon of overshadowing is in agreement with the concurrent interference principle. During training, a strong CS and a weak CS were presented simultaneously prior to a UCS until the compound reliably elicited the CR. During extinction tests, the weak CS and the strong CS were each presented alone on different trials. It was found that the strong CS presented by itself elicited the CR during extinction, but the weak CS did not. Pavlov's own description of this phenomenon, although certainly not written in terms of the present approach, shows how overshadowing might be considered a type of concurrent interference: "In the usual case where two hitherto neutral stimuli are used to form a compound stimulus, the stronger stimulus at once prevents the weaker from forming a corresponding connection with the center for the unconditioned reflex (Pavlov, 1927, p. 144)." Table I outlines the

Table I

Overshadowing Considered Operationally and in Terms of the Interference Rule

	Operational	Theoretical
Training	Overshadowing: (weak CS and strong CS)→UCS Control: weak CS→UCS	Overshadowing: (weak E-pre and strong E-pre-X)→E-post Control: weak E-pre→E-post
Test	The weak CS is less likely to elicit a CR under the overshadowing procedure than under the control procedure	Because E-post becomes associated with the strong E-pre-X under the overshadowing procedure, the probability of an association of E-post with the weak E-pre is reduced

overshadowing experiment in the column labeled "operational" and translates his description of it into the language used here in the column labeled "theoretical." The weak CS is called E-pre because the concern is with its association with the UCS (E-post). The strong CS is called E-pre-X because the concern is with the interference it may produce with the reference association. Since E-pre-X was stronger than E-pre, it became associated with E-post more rapidly and concurrently interfered with the reference association. Note that the present formulation does not imply that there was no association between E-pre and E-post but simply that the magnitude of the association was reduced; more subtle experimental probes show that the animal exhibits a reduced association of E-pre with E-post (Baker, 1968; Pavlov, 1927).

   b. Examples from Discrimination Learning. In discrimination learning, the concern typically is with the control of behavior by a stimulus dimension rather than by particular stimuli. The speed with which a dimension comes under stimulus control is called its salience and is similar to what we have called associative strength. Honig (1970) has summarized a great deal of data from experiments in which an animal is subjected to discrimination training involving two redundant sets of discriminative stimuli so that effective performance can occur if the discrimination is based on only one of the sets. In such experiments the presence of a more salient stimulus dimension tends to prevent stimulus control by the less salient dimension. Thus overshadowing occurs in discrimination learning. Furthermore, it is tenable to suppose that the mechanism underlying this overshadowing is concurrent interference. Sutherland and Mackintosh (1964) trained rats on a discrimination problem involving two redundant cues, called A and B, and then tested each cue separately. The more cue A controlled behavior, the less cue B controlled behavior and vice versa. Reynolds (1961) obtained a similar result with pigeons. Although these results have been explained in terms of attention theory, they are compatible with the hypothesis that the association of one cue with the consequences of responding tends to prevent the other cue from becoming associated.

2. Prior Experience with E-pre-X

According to the concurrent interference principle, there are two major ways in which prior experience with E-pre-X can change the amount of interference it produces.

1. Suppose E-pre-X has already become associated with E-post prior to training. During training, E-pre is introduced so that both E-pre and E-pre-X precede E-post. According to the concurrent interference principle, E-pre-X will tend to produce more interference than if it were novel because it is already associated with E-post.

2. On the other hand, suppose E-pre-X has been repeatedly presented in the absence of E-post prior to training. There is ample evidence from the classical conditioning literature that this prior exposure will reduce the associative strength of E-pre-X relative to E-post (Carlton & Vogel, 1967; Lubow, 1965; Rescorla, 1969). If such an E-pre-X of reduced associative strength occurs during training, the concurrent interference principle predicts that it will produce less interference than if it were novel.

a. Examples from Classical Conditioning. Kamin (1969) obtained results that are compatible with the first of these deductions, that pairing E-pre-X with E-post prior to training will increase the amount of interference produced by E-pre-X. He called this phenomenon "blocking." In a conditioned suppression situation, rats in a blocking group were pretrained to associate a noise (E-pre-X) with shock (E-post); for a control group, this pretraining phase was omitted. During training, a light (E-pre) and the noise (E-pre-X) were simultaneously presented prior to shock. During a later test in which the light (E-pre) alone was presented, the blocking group did not exhibit conditioned suppression, but the control group did. This result indicates that the prior association of the noise and the shock resulted in increased interference by the noise.

Regrettably, I know of no findings from conventional classical conditioning experiments which can be used to support the second prediction, that interference produced by E-pre-X is reduced when earlier repeated occurrence of

E-pre-X in the absence of E-post has reduced its associative strength. In Section VI, D of this chapter, I will report experiments from my laboratory which support this prediction; they are based on the association of flavors with toxicosis.

b. Examples from Discrimination Learning. The following passage by Hilgard and Bower (1966, p. 532) shows that a counterpart of Kamin's blocking effect occurs in discrimination learning:

> ...A trained bias toward using a particular type of cue (call it cue A) can be made to show itself by comparison with control ("unbiased") subjects during the learning of a new problem, and by one of three methods: (1) facilitation of learning when the relevant cue is still of type A, (2) interference with learning when a different cue is relevant whereas cue A, although present, is now irrelevant to the correct solution, and (3) a preference for learning mainly about cue A when cue A and a different cue, B, are made redundant and equally relevant in the new problem.

If the blocking effect which occurs in discrimination learning can be attributed to concurrent interference, the blocking of dimension B by dimension A must occur because the animal has already learned to form a discrimination based on dimension A and continues to do so. Mackintosh and Honig (1970) have supplied evidence compatible with this deduction, even though they interpret their results in terms of selective attention. They point out, however, that there are circumstances in which the blocking effect can be overriden by another factor: The general tendency of past discrimination training to improve performance on subsequent discrimination problems, even those involving novel dimensions.

3. Relevance of E-pre and E-pre-X to E-post

"Relevance" is roughly equivalent to Thorndike's law of belongingness (Hilgard & Bower, 1966, p. 28). In the present terminology, it means that one cannot always determine the associative strength of an E-pre without knowing the type of E-post with which it is to become associated. For instance, the associative strength of a flavor is high when E-post is toxicosis and is low when E-post is an electrical shock; the associative strength of an telereceptive or proprioceptive stimulus is low when E-post is toxicosis and is high when E-post is electrical shock. "Relevance" is simply a term used to refer to such relationships. Its implication in the context of concurrent interference theory is that if the events which occur during a delay are relevant neither to E-pre nor to E-post, they will not produce interference with the reference association, particularly when E-pre and E-post are highly relevant to each other. Evidence for these assertions will be deferred until Section V, when relevance will be discussed in more detail and will be used to explain those spectacular instances of long delay learning mentioned in the introduction to this chapter.

D. The Second Type of Concurrent Interference: A Leap of Faith

The preceding section contains substantial evidence for interference produced by association of E-pre-X with E-post, which was called the first type of concurrent interference. There is no experimental evidence for the second type, which is produced by the association of E-pre with E-post-X. It is supposed here that methodological problems are responsible for this lack of evidence. As an example, consider classical conditioning. To demonstrate the first type of concurrent interference is easy. A suitable E-pre and a suitable E-pre-X can readily be selected and paired with E-post. During a test, each can be presented separately and the strength of each association with E-post can be measured by the magnitude of the CR. To demonstrate the second type, E-pre must precede an E-post and an E-post-X and, during the test, it must be determined how much of the animal's reaction to E-pre is to be attributed to an association with E-post and how much is to be attributed to an association with E-post-X. The minimum condition for such a test to be successful is that each association must produce a different CR and that these CR's should not interact with each other in any important way. Since it is hard to think of two CR's which might not conceivably interact, this minimum condition imposes nearly insurmountable technical difficulties. An instrumental learning situation would pose even more severe difficulties.

A consideration of symmetry led to the postulation of the second type of concurrent interference in spite of the difficulty of obtaining direct evidence in its favor. If the association of an extraneous event with one element of the reference association produces interference, a similar association with the other element should also produce interference. In other words, since an association between E-pre-X and E-post produces interference, an association of E-pre with E-post-X should also produce interference by a principle of symmetry. Admittedly, this is merely an intuitive consideration and hardly meets the criteria for evidence. Thus it is important to note that a version of concurrent interference theory could still be viable without the second type of concurrent interference. (The most important change necessary in such a version of the theory would be in Section III, F, where a proximity corollary is derived partly on the assumption that the second type of concurrent interference exists. This proximity corollary would have to be considered an independent postulate if the second type of concurrent interference was not presupposed.)

E. Concurrent Interference vs. Selective Attention

A number of experimental effects from the classical conditioning literature and from the discrimination learning literature have been shown to be explicable in terms of the first type of concurrent interference in Section III, C. Unfortunately, the same effects can also be explained in terms of selective attention theory. The rather subtle differences between concurrent interference and selective attention will be delineated in this section.

The similarity between selective attention and concurrent interference is that both imply an inverse relationship between the association of E-pre with E-post and the association of E-pre-X with E-post. According to the version of attention theory to be used here for expository purposes, the more attention an animal pays to E-pre-X, the less attention it can pay to E-pre and vice versa; of course, the attention paid to E-pre or E-pre-X is presumed to be a determinant of whether it will become associated with E-post. Attention theory accounts for the overshadowing phenomenon on the basis that the attention paid to a strong CS prevents attention from being paid to a weak CS. It accounts for the effects of past experience on the amount of interference produced by E-pre-X with the hypothesis that experience changes the amount of attention the animal pays to E-pre-X.

The difference between the attention approach and the concurrent interference approach advocated here may be illustrated by the difference in the way each interprets Kamin's blocking effect. It will be recalled that Kamin paired an E-pre-X (noise) with E-post (shock) during a pretraining phase. During training, E-pre (a light) was presented together with E-pre-X prior to E-post. E-pre did not become associated with E-post although with the omission of the pretraining phase or with the omission of E-pre-X during training, it did become associated.

According to a straightforward attention theory, blocking occurred because the pretraining experience insured that the animal would attend to E-pre-X during training and hence would not be able to attend E-pre. According to concurrent interference theory, blocking occurred because the association between E-pre-X and E-post prevented the association between E-pre and E-post from developing. The two theories yield different predictions for a modification of the blocking procedure in which the nature of E-post is changed from pretraining to training. Suppose E-pre-X is associated with food during pretraining instead of shock and that during training the shock is substituted for food. Attention theory implies that blocking will still occur because the animal has learned to pay attention to E-pre-X. Concurrent interference theory implies that the change in the nature of E-post from food to shock will prevent blocking. Thus blocking should occur when shock is used throughout because E-pre-X was associated with shock, but it should not occur when E-pre-X has been associated with food instead of shock.

A major reason for the author's enthusiasm for this particular example is that Kamin (1969, pp. 57–63) has found that the blocking effect was substantially attenuated when the intensity of shock was increased from pretraining to training or when a pair of shocks was administered instead of a single shock. Insofar as blocking was attenuated by the change in the nature of E-post,

Kamin's results are compatible with the concurrent interference interpretation. However, strictly speaking, the change in E-post should have prevented blocking altogether according to concurrent interference theory. It is contended here, that the reason blocking was not eliminated completely was that the change in the nature of E-post was very slight: it was only a change in the nature of the shock. It would be expected that when the E-post used during pretraining is very different from that used during training, no blocking at all would be obtained.

For expository purposes, attention theory has been given an explicit form here which may be naive. I apologize to those whose more sophisticated work has been ignored. Furthermore, it should be emphasized that attention and concurrent interference need not be mutually incompatible processes. There is no reason both processes cannot occur simultaneously. However, Kamin's results are generally more compatible with the concurrent interference approach than with attention theory, and Sections V and VI of this chapter will describe results which cannot be explained by attention but are easily explained by concurrent interference. Thus when a result can be explained in terms of either theory, it should not be automatically assumed that the attention approach is preferable simply because it has been more popular in the recent past.

F. The Proximity Corollary

Compare the case in which E-pre and E-post occur without a delay between them with the case in which they occur with a substantial delay between them. It has already been suggested that the delay results in more interference which, in turn, results in poor learning of the reference association. However, this is not a complete explanation. Since we consider the world a place in which potentially interfering events occur sporadically in time and independently of E-pre and E-post, a delay period should not produce an increase in the number of interfering events but only a shift in when they occur. The interfering events, which occur between E-pre and E-post in the delay case, must occur either prior to E-pre or after E-post in the no-delay case. Consider a concrete example: In the delay case, E-pre might occur at 1:00 PM and E-post might occur at 1:10 PM, while in the no-delay case, both events occur at 1:05 PM. In the delay case, the interfering events which occur between 1:00 and 1:10 occur between E-pre and E-post, while in the no-delay case, they occur either before E-pre or after E-post. Thus in order for the proposed approach to account for the effect of an increased delay rigorously, the following proximity corollary must be derivable from the concurrent interference principle. *Events which occur between E-pre and E-post produce more interference than events which occur prior to E-pre or after E-post.*

1. Derivation

In the present theoretical context, the terms "E-pre-X" and "E-post-X" describe the functions of events rather than the natures of the events themselves. An event is called "E-pre-X" when the concern is with the interference it produces by becoming associated with E-post. The same event may be called "E-post-X" when the concern is with interference produced by a potential association with E-pre. In the derivation of the proximity corollary, it is convenient simply to refer to events and discuss their roles as E-pre-X's and E-post-X's separately. The corollary will be verified if the probability of interference is greater when a potentially interfering event, regardless of its role, occurs during the delay between E-pre and E-post than when it occurs either (1) prior to E-pre or (2) after E-post.

Begin by considering events in their roles as E-pre-X's. If an E-pre-X occurs prior to E-pre, it is less likely to produce interference than if it occurs during the delay because moving E-pre-X back earlier in time increases the number of events which intervene between it and E-post. As these intervening events run off in sequence, there is an opportunity for E-pre-X to become associated with each of them, reducing its likelihood of becoming associated with E-post when E-post finally occurs. If an event occurs after E-post instead of during the delay, it can no longer produce interference in its role as E-pre-X because it can no longer become associated with E-post in a forward direction.

Now consider the role of events as E-post-X's. If an E-post-X occurs prior to E-pre instead of during the delay, it cannot produce interference because it cannot become associated with E-pre in a forward direction. If it occurs after E-post, then E-pre has already had an opportunity to become associated with E-post before E-post-X occurs. An association of E-pre with E-post would reduce the strength of the association between E-pre and E-post-X and hence reduce the amount of interference E-post X produces. (It may seem paradoxical that interference with an interfering association can be produced by the reference association. However, it should be remembered that the events of the reference association are distinguished from other events only in that they have been selected for observation by the experimenter. Thus the same principle of concurrent interference applies to them also.)

2. The Proximity Corollary and Temporal Contiguity

The proximity corollary implies the following subcorollary, which shows that concurrent interference theory makes the same predictions as the law of temporal contiguity in many situations: *Among a number of E-pre's and E-pre-X's, which otherwise would be equally likely to become associated with E-post, those closer in time to E-post are more likely to become associated with E-post.*

The derivation of this subcorollary is as follows: Suppose two E-pre's, designated as E-pre-1 and E-pre-2, occur prior to E-post with E-pre-1 occurring first so that E-pre-2 is closer in time to E-post. If both E-pre's are of equal associative strength relative to E-post, the E-pre which is subjected to the greater amount of interference will be the more weakly associated. According to the proximity corollary, more interference is produced by an event which occurs during the delay between E-pre and E-post than by an event which occurs prior to E-pre. Thus the interference must be greater for the association between E-pre-1 and E-post than for the association between E-pre-2 and E-post, since E-pre-2 occurs during the delay between E-pre-1 and E-post, while E-pre-1 occurs prior to the delay between E-pre-2 and E-post. The net result is that E-pre-2, the event occurring closer in time to E-post will be more strongly associated.

The subcorollary shows that for those situations in which many events capable of becoming associated with E-post occur, concurrent interference theory has about the same behavioral implications as the traditional hypothesis that temporal contiguity (or near contiguity) is necessary for learned associations. This should come as no surprise. The theory was designed to explain the frequent failures to obtain associations over a delay without the presupposition that associative memory is in principle short. The main difference between concurrent interference theory and more traditional approaches lies in the fact that the proximity corollary and its subcorollary are both rules about the interaction of a reference association and other potential associations. If there are no other potential associations, these rules are not applicable; and if the other potential associations are very weak, they become of minor importance. In this way, concurrent inteference theory allows for association over long delays when the delay does not include events which can become strongly associated either with E-pre or with E-post.

IV. Delay of Reinforcement in Instrumental Learning

In the terminology used here, delay of reinforcement refers to the case in which E-pre is a response, E-post is a reward (or punishment), and there is a delay between them. The problem is whether the occurrence of the reward (E-post) will increase the probability of the response (E-pre). The proposed view is that the detrimental effects of the delay on learning occur because the delay includes interfering events. In the case of the manipulative and running behaviors usually studied in delay of reinforcement experiments, interference is quite likely to occur. Even if all exteroceptive stimuli were to be eliminated from the delay, the animal would still emit movements which would produce interference.

A. RESPONSE COMPETITION VS. CONCURRENT INTERFERENCE

At a theoretical level, the proposed view is diametrically opposed to the 1947 position of Spence who maintained that, without secondary reinforcement, learning is impossible even after a short delay of reinforcement. Remarkably enough, in the analysis of the role of delayed reinforcement in specific experimental situations, Spence's position is so markedly similar to the present position that it requires some subtlety to tell them apart. Consider Spence's description (1956, p. 154) of what may happen during a delay to reduce the efficacy of reinforcement:

> Remaining as it does in the situation the animal now makes other responses; thus it may turn away from the loci of the lever, it may run around in the runway or the goalbox, it may attempt to jump up to the top of the box, bite at the door, and so on. Occurring as these responses do to essentially the same stimulus components as does the to-be-learned instrumental response, they likewise become conditioned to them. Thus interfering responses are established which, insofar as they tend to occur on a training trial, will have the effect of increasing the time it will take the appropriate response sequence to run off. With immediate reinforcement, on the other hand, these competing responses are not established and hence there would be little or no interference.

Although there is a difference between Spence's 1956 position and the proposed position, note the remarkable similarity. Spence claims that the uncontrolled responses occurring during the delay are responsible for the reduced effectiveness of delayed reinforcement. The proposed view is that all events, whether they be responses or exteroceptive stimuli, can produce interference. However, it cannot be denied that in practice, responses are the most important of these interfering events because there is substantial experimental control of exteroceptive stimulation. Indeed, Spence's 1956 position is more compatible with the present view that associative memory is long than with his own 1947 view that it is exceedingly short; for if the latter is really true, then there is no reason to invoke competing responses to explain why delayed reinforcement has reduced effectiveness. Moreover, in his 1956 analysis of delayed reinforcement, Spence made no specific reference to temporal contiguity theory, although, conceivably, a weakened version may be present in a tacit assumption that intervening responses are more strongly reinforced than the reference response. If so, this weakened version of temporal contiguity theory is similar to the present proximity corollary.

The main difference between Spence's competing response theory and concurrent interference theory is the mechanism by which responses occurring during the delay reduce the effectiveness of delayed reinforcement. The competing response theory holds that the reinforcement increases the probability of the intervening behaviors. Because a number of responses cannot occur simultaneously, the probability of the reference response is reduced. Concurrent interference theory presupposes, on the basis of the proximity corollary, that the intervening behaviors will tend, all other things being equal, to become more strongly associated with the reinforcement than the reference response. Hence, they will interfere with the association of the reference response with reinforcement.

In their interpretations of the delayed reinforcement of manipulative and running responses, these theories differ so subtly that only the most ingenious experimentation could possibly distinguish between them. There are two reasons such a distinction is very difficult: (1) In the types of experiments under consideration, unless special precautions are taken, response competition and concurrent interference can occur simultaneously. Thus the two approaches are not usually incompatible. (2) Results which are accounted for by response competition usually are automatically accounted for by concurrent interference; the same increase in the probability of the intervening responses which is responsible for the reduced effectiveness of delayed reinforcement according to Spence, is evidence, according to concurrent interference theory, that the intervening responses became associated with the reinforcer and hence produced concurrent interference. Consequently, the data cited by Spence (1956, pp. 153–164) in support of his theory are also compatible with the theory proposed here.

There is, however, a difference in flavor between the response competition theory and concurrent interference. To illustrate this, I will comment on Spence's analysis of some experiments in which depression of a lever by a rat results in immediate withdrawal of the lever followed by delayed food reinforcement (presented near the lever). A shift from a 1.0-second delay to a 10.0-second delay of reinforcement did not markedly disrupt performance in an experiment by Harker (1956). In an experiment by Shilling (1951), a shift from a 1.0-second delay to a 5.0-second delay had little effect on performance, although a shift to 10.0 seconds produced some decrement. In both of these cases, the shift from a 1.0-second delay to a longer delay resulted in better performance than that obtained when the animal started with a longer delay of reinforcement. Observations by both Harker and Shilling indicated that the effect of delayed reinforcement was correlated with the rat's maintenance of its orientation toward the lever during the delay. If the rat failed to remain in the vicinity of the lever during the delay, the delayed reinforcement was less effective. This is how Spence (1956, p. 160) explained this effect:

> The theoretical implication of these observations of Harker and Shilling is, I think, obvious. They suggest the hypothesis that the decrement in performance with delay of reinforcement is contingent, in part at least, on whether the subject maintains an orientation toward the stimulus complex of the response manipulandum and food box during the delay period or whether it gives up this response adjustment and substitutes other conflicting ones, such as turning away. Their findings suggest that there is a limit to which this orientation can be maintained in the rat, which is a function, in part, of the type of situation employed. Thus Harker's subjects successfully maintained the orientation for 10 seconds and performed at a level equal to that of subjects that had a 1-second

delay. In the type of situation Shilling employed, on the other hand, the animals were not able to maintain their orientation for 10 seconds but could for 5 seconds.

The concurrent interference approach shares Spence's view that if rats remain oriented toward the lever and food box during the delay, fewer interfering behaviors can occur, and hence delayed reinforcement will be facilitated. However, Spence does not explain why this orientation is maintained if rats are switched from immediate to delayed reinforcement and not if training begins with delayed reinforcement. The concurrent interference approach can account for the development of this orientation, admittedly on an *ex post facto* basis. When training begins with delayed reinforcement, the behaviors which occur during the delay produce concurrent interference and prevent the reference association from occurring. However, if the rat is pretrained with nearly immediate reinforcement, it learns an association between the lever press and the reinforcement prior to training with delayed reinforcement. We have seen from Kamin's blocking effect that such an association developed prior to training will tend to prevent other associations from occurring during training. If so, the prior association of lever-pressing with reinforcement might be expected to interfere with the association of the intervening behaviors with reinforcement. Thus the intervening behaviors will tend to drop out and the animal will remain oriented toward the lever and the food box. In practice, the amount of delay which can still result in maintenance of performance will depend upon the strength of the association of the lever press with reinforcement relative to the total strength of the associations between intervening responses and the reinforcement. Since the total amount of interference produced by interfering events increases as the length of the delay increases, there must a delay at which the prior association of the lever press and the reinforcement is no longer able to to block the interference; at that delay, the concurrent interference might be expected to drown out the association between the lever press and the reinforcement instead of itself being drowned out. This point was less than 10 seconds in the Shilling experiment and beyond 10 seconds in the Harker experiment. At its present level of development, the present approach cannot account for this difference and, as far as can be discerned from his 1956 book, neither could Spence. Perhaps eventually a closer analysis of the two experiments might account for this difference in their results.

Since response competition and concurrent interference are not incompatible processes, it is conceivable that delayed reinforcement situations exist in which response competition has a more important role than concurrent interference. But while I know of no such situations, there certainly are many instances of associative memory to which concurrent interference theory is applicable and to which response competition theory is not applicable. These include the effect of the duration of the interstimulus interval in classical conditioning, as well as the Garcia and Capaldi demonstrations of long-delay learning. Thus the main advantage of concurrent interference theory over response competition theory is its greater generality.

Despite the explanatory power and unusual generality of concurrent interference, there is likely to be a prejudice against it which I would like to combat. The prejudice stems from the historical bias of American behaviorism in favor of peripheral explanatory mechanisms, like response competition. Both Hullians and Skinnerians have preferred to postulate peripheral mediating reactions, such as those occurring in the salivary glands or in the muscles, than to postulate central mediating reactions which occur in some *terra incognita* such as the nervous system. Both types of reactions involve biological systems, but the peripheral reactions have the advantage of involving better known biological mechanisms; they are easier to measure, easier to deal with in terms of conditioning paradigms, and less prone to be used as synonyms for mentalistic processes. Response competition shares these advantages, while interference with a reference association does not. Nevertheless, it now seems fair to claim that the peripheralistic approach has not been so notably successful that it can justify the sacrifice of the greater generality of the concurrent interference approach.

B. THE SECONDARY REINFORCEMENT PARADOX

Secondary reinforcement which occurs during the delay between a response and a primary reinforcement is known to increase the efficacy of delayed reinforcement. This effect of secondary reinforcement is an apparent paradox for concurrent interference theory because, strictly speaking, the net effect of a secondary reinforcer must be to produce interference. In the present terminology, the response is E-pre and the primary reinforcer is E-post; relative to E-pre, the secondary reinforcer is E-post-X and relative to E-post, it is E-pre-X. If so, a secondary reinforcer results in two interfering associations and thus its occurrence should interfere with the reference association. Obviously, this suggests that a secondary reinforcer should also interfere with the efficacy of delayed reinforcement.

The proposed explanation of the facilitation of delayed reinforcement by secondary reinforcement begins with the acceptance of secondary reinforcement itself as an empirical fact which need not be explained. That is, it is assumed that a neutral stimulus which has become associated with a primary reinforcer is capable of itself acting as a reinforcer for some new response (Kelleher, 1966; Kelleher & Gollub, 1962). This assumption enables a resolution of the apparent paradox already described; for then the presence of a secondary reinforcer can facilitate delayed reinforcement without also facilitating the development of the reference association. The secondary reinforcer acts as a reinforcer of the response in its own right, while its association with the primary reinforcer maintains its ability to act as a reinforcer.

The first implication of this interpretation is not readily testable; secondary reinforcement will facilitate delayed reinforcement only if there is a potential for interference with the reference association. If, however, there are no other events during the delay, the only result of the secondary reinforcer should be interference with the reference association so that the response is associated with the weaker secondary reinforcer rather than with the primary reinforcer.

A second implication of the proposed theory is more striking. Secondary reinforcement need occur only during two portions of the delay in order to facilitate delayed reinforcement. It must follow the response in order to act as a reinforcer of the response and it must precede the primary reinforcer in order to retain its power to reinforce; during the remainder of the interval, it need not be present at all. Thus if there is a 30-second delay of primary reinforcement and the secondary reinforcer is a light, presumably the light can potentiate delayed reinforcement, if it is present only for a short time following the response, a short time preceding the primary reinforcer, and is absent for the remainder of the interval. Indeed, the absence of the light during the remainder of the interval might actually increase the effectiveness of delayed reinforcement, for Bersh (1951) has found that the presence of a light for a prolonged period prior to the primary reinforcer may actually weaken its association with the primary reinforcer. (Conceivably, however, this advantage of extinguishing the light during the middle of the delay might be balanced by disruption resulting from the extra light offset and onset in the midst of the delay.)

Suggestive evidence that the secondary reinforcer need not be present during the entire delay is supplied by the success of Ferster and Hammer (1965) in using the color of a panel as a secondary reinforcer to obtain effective reinforcement with a 24-hour delay of primary reinforcement in a lever-pressing situation with primates. Surely the secondary reinforcer must have been effectively interrupted during the delay if only because primates cannot be looking at panels while they sleep. Furthermore, C. C. Perkins, Jr. has pointed out to me that an experiment by Bixenstine (1956) also suggests that the secondary reinforcer need not be present during the entire delay. The experiment involved the association of a discriminative stimulus with delayed punishment of eating, rather than delay of positive reinforcement. For all groups, presentation of a blinking light as a rat was approaching a food trough indicated that eating would be punished with electrical shock after a delay. For the experimental groups, the blinking light also preceded the shock by 3 seconds; not so for the control groups. The different experimental and control groups varied in the delay of punishment from about 20 seconds to about 110 seconds. All the experimental groups learned not to eat in the presence of the blinking light about equally well despite the great differences in delay of punishment. For the control groups, the punishment was ineffective at the 110 second delay. Thus the blinking light was able to facilitate the delayed punishment even though it was absent during nearly the entire delay.

V. Relevance

According to concurrent interference theory, the detrimental effect of a delay on the association between an E-pre and an E-post results from interference produced by events occurring during the delay. Conversely, when an association develops between an E-pre and a long delayed E-post, there must be relatively little interference by intervening events. Since a very long delay (i.e., of hours) contains a large number of events capable of entering into associations, these events must somehow be prevented from interfering with the reference association. The characteristic of associative learning which prevents this interference from occurring is relevance, which has already been described briefly, but now will be considered in more detail.

In order for the concurrent interference approach to explain long-delay learning, relevance must insure that the association between E-pre and E-post is stronger than the combined strength of all interfering associations. This suggests strongly that the following four conditions must hold if long-delay learning is to occur:

1. E-pre and E-post must have high associative strength relative to each other.

2. In their roles as E-pre-X's, the intervening events must have low associative strength relative to E-post. If not, the E-pre-X's would tend to become associated with E-post and prevent the reference association from occurring.

3. E-pre must have low associative strength relative to the intervening events in their roles as E-post-X's. If not, E-pre will have become associated with a number of E-post-X's at the time E-post finally occurs.

4. The intervening events must have high associative strength relative to each other. That is, intervening events in their roles as E-pre-X's must readily become associated with other intervening events functioning as E-post-X's. Such associations would reduce the number of intervening events available to become associated with E-pre or E-post.

A. Stimulus Relevance

Stimulus relevance is the type of relevance responsible for delayed flavor-toxicosis associations. When the above four conditions for long-delay learning are applied to learned flavor aversions, they can be specified as follows:

1. Flavors and toxicosis must readily become associated.

2. Intervening events, in their roles as E-pre-X's, must not readily become associated with the toxicosis. Strictly speaking, an experimental demonstration that this condition holds is not feasible because there are bound to be a huge number of intervening events during a long delay, some of which are difficult to control experimentally. Presumably, however, the vast majority of these emanate directly from the external environment or are movements of the animal

itself. Thus, the testable implications of this assertion are that it is hard for an animal to learn that a sound, a light, or a self-produced movement will make it sick.

3. Flavors must not readily become associated with the intervening events in their roles as E-post-X's. The testable implications of this assumption depend on those intervening events which appear to emanate directly from the external environment. Presumably, it is hard for an animal to learn that consumption of a particular flavored substance results in the occurrence of a particular environmental outcome.

4. The intervening events must readily become associated with each other. The testable implication of this assumption is that telereceptive stimuli and movements of the animal should readily become associated with environmental outcomes.

1. Indirect Evidence

Garcia and his co-workers have supplied direct evidence that each of the four conditions necessary for associations of flavors with delayed toxicosis actually exists. This chapter would be logically complete with only this evidence for stimulus relevance without any attempt to explain the underlying mechanism since it is my aim to use the concept of relevance to explain long-delay learning rather than to explain relevance itself. Nevertheless, this chapter would not be intuitively satisfying without some consideration of the nature of stimulus relevance prior to a summary of evidence that it exists.

Garcia and Ervin (1968) hypothesized the existence of two nearly separate systems in associative learning: (1) the internal system in which gustatory stimuli readily become associated with visceral aftereffects and (2) the external system in which telereceptive stimuli readily become associated with cutaneous aftereffects. According to Garcia and Ervin, it is very difficult to produce an association of events in one system with events in the other system. Obviously two such systems would result in the four conditions necessary if flavors are to become associated with delayed toxicosis.

Garcia and Ervin (1968) supply some suggestive evidence that the brain is anatomically so organized as to result in these two systems for associative learning. Furthermore, there are evolutionary grounds to expect development of two such systems for association. It is adaptive for animals to associate differentially the flavors of ingested substances with long delayed toxicosis or other long delayed physiological aftereffects for two reasons: (1) under natural conditions an ingested substance only affects the internal environment of the animal, and (2) these internal aftereffects often do not occur until hours after ingestion. Even if the proposed concurrent interference theory is not accepted in detail, it still is difficult to imagine how an animal born with a *tabula rasa* would be able to regulate its food intake on the basis of prior experience with delayed visceral aftereffects given the large number of stimuli bound to occur during a delay. This alone is strong presumptive evidence that something like the two systems hypothesized by Garcia and Ervin (1968) must exist. However, it will become apparent later that recent findings cast doubt upon the exact scheme that they hypothesized.

2. Bright, Noisy Water

To show that there are separate internal and external systems for associative learning, Garcia and Koelling (1966) selected an E-pre and an E-post from each system in order to show (1) that an association would occur readily if E-pre and E-post were from the same system and (2) that an association would not occur readily if E-pre and E-post were from different systems. More specifically, Garcia and Koelling used a discrimination learning paradigm in which a drinking response was punished if it occurred in the presence of a discriminative stimulus and then determined if that discriminative stimulus inhibited drinking. The discriminative stimulus is designated as E-pre and the punishment is designated as E-post, while the drinking itself is given no particular designation because it occurred in all the experimental procedures.

The internal E-pre was the flavor of saccharin in the water and the internal E-post was delayed toxicosis produced by X-irradiation. The external E-pre was a light–sound combination and the external E-post was immediate electrical shock. Electrical shock was presumed to be in the external system because it produces peripheral pain rather than internal discomfort and also because if two rats in a single chamber are shocked, they begin to fight each other (Ulrich & Azrin, 1962) just as if each rat attributed the pain to the other.

Garcia and Koelling (1966) used a noteworthy methodological innovation to make the light–sound combination take on the role relative to drinking that flavors usually have; it is called bright, noisy water. A drinkometer circuit was used to produce an audible click and a flash of light whenever rats licked tap water at a spout. Thus telereceptive stimulation was produced with each lick much as flavors become intensified with each lick.

Figure 1 shows that Garcia and Koelling obtained results which support the doctrine of stimulus relevance as follows: (1) Punishment of consumption of saccharin solution by X-irradiation was effective. (2) Punishment of consumption of saccharin solution by shock was ineffective. (3) Punishment of consumption of bright, noisy water by X-irradiation was ineffective. (4) Punishment of consumption of bright, noisy water by shock was effective. Thus Fig. 1 is evidence in favor of the four conditions which are necessary if the concurrent interference theory is to account for delayed flavor–toxicosis associations.

Garcia and Koelling (1967) further confirmed this result with a number of other experimental procedures. They also showed that smells have associative

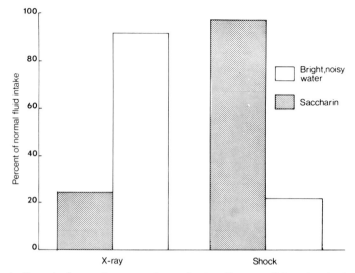

Fig. 1. Percent of normal consumption under two E-pre conditions (saccharin flavored water and bright, noisy water) as a function of the type of E-post (X-irradiation or electrical shock) with which E-pre had previously been paired. A low ordinate value indicates the occurrence of an association. (Data courtesy of J. Garcia and R. A. Koelling.)

properties intermediate between those of flavors and exteroceptive stimuli. That is, smells become associated with toxicosis more readily than telereceptive stimuli and less readily than flavors; when the punishment is electrical shock, the converse is true. Finally, Braveman and Capretta (1965) and Dietz and Capretta (1967) have also shown that flavors become associated more readily with toxicosis than with shock. It should be noted, however, that none of these experiments involved extensive training. With more training, it is possible to develop an apparent association between telereceptive stimuli and immediate toxicosis as well as between flavors and immediate shock (Revusky & Garcia, 1970, pp. 26–31).

3. Big vs. Little Pellets

Garcia, McGowan, Ervin, and Koelling (1968) have supplied suggestive evidence that the movements of the animal have associative properties similar to those of telereceptive stimuli; that is, they have high associative strength relative to shock and low associative strength relative to toxicosis. One group of rats was shocked after seizing large pellets and a second group was shocked after seizing small pellets. The rats learned to avoid the size of pellet associated with shock. There are two important differences between large and small pellets: the difference in appearance and the different ways in which the rats must seize them. The rats were able to associate these differences with shock. However, when the punishment was toxicosis, it was not possible to produce avoidance of the punished size of pellet. This indicated that neither the appearance of the pellets nor the proprioceptive stimuli concurrent with their seizure could become associated with toxicosis. Of course, it should not come as a surprise that control experiments showed that toxicosis could become associated with differences in flavor but not with differences in size.

More experiments about stimulus relevance are summarized by Revusky and Garcia (1970, pp. 26–31, 41–43). Note, in passing, that all such results are not explicable in terms of attention theory; for if the associative strength of an E-pre depended only on the animal's attention to it, then the nature of E-post would not matter.

4. Color Cues

Capretta (1961) showed that chickens can associate the color of mash with concurrent toxicosis. This contradicts the two-systems approach of Garcia and Ervin (1968) since color should be in the external system and hence irrelevant to toxicosis. Until recently, I felt that the ability to associate colors with toxicosis might be a secondary evolutionary mechanism limited to birds; presumably, it might have developed because birds often must select food primarily on the basis of color. Recently, however, evidence has been found that associations of the color of an ingested substance with physiological consequences may have a more important role in the regulation of food intake than previously supposed.

Wilcoxon, Dragoin, and Kral (1969) showed that quail, like rats, could develop an aversion to sour, uncolored water if its consumption were followed by toxicosis induced by injection of cyclophosamide 30 minutes later. The quail also could develop an aversion to unflavored, blue water under similar circumstances. Finally, the experimenters tried to determine whether color or flavor was the more important cue for the birds. During conditioning, quail consumed water which was both sour and blue and then were subjected to toxicosis. A subsequent test revealed a pronounced aversion to unflavored blue water but only a slight aversion to uncolored sour water. Thus the color cue actually appeared to be dominant over the flavor cue at the particular stimulus intensities used.

Gorry and Ober (1970) showed that squirrel monkeys also can associate delayed toxicosis both with the flavor and with the color of consumed substances. An experimental group was allowed to drink green sucrose solution and was injected with lithium chloride solution 30 minutes later in order to produce toxicosis. A familiarization control group was injected with harmless saline solution instead of lithium chloride following consumption of green sucrose solution. A sensitization control group consumed green sucrose solution on the day prior to conditioning; on the day of conditioning, consumption of uncolored and unflavored water was followed by lithium chloride toxicosis.

The first test day occurred three days after training. All the monkeys were given a choice between uncolored unflavored water and a test substance. For half the monkeys in each group, the test substance was unflavored green water; for the other half, it was uncolored, sucrose solution. The experimental treatment produced an aversion both to the green color and to the sweet flavor of sucrose relative to each of the control treatments.

On the following day, there was a second test in which the monkeys which had previously been tested with sweet water were tested with green water and vice versa. The learned aversion to the sweet water was still exhibited by the experimental monkeys, but the aversion to the green color had disappeared.

On the following day, there was a third test designed to determine which cue, flavor or color, was stronger relative to toxicosis. All monkeys were given a choice between unflavored green water and uncolored sweet water. The experimental monkeys drank the green water in preference to the sweet water, suggesting that the color cue was far less potent than the flavor cue Note, however, that it is not certain that the same result would have been obtained if this test had been the first of the series.

The preceding experiment did not definitively demonstrate that monkeys can associate the color of a consumed substance directly with toxicosis. Conceivably, the monkeys might have associated only the flavor with the delayed toxicosis and avoided the green water because the green color had been associated with the punished flavor (see Capretta & Moore, 1970). If so, monkeys should be unable to associate unflavored colored water with delayed toxicosis. Gorry and Ober excluded this possibility in an experiment with the control animals of the preceding experiment. Half the monkeys received red water on Day 1, yellow water followed 30 minutes later by toxicosis on Day 2, and daily choice tests between the colors beginning on Day 5. For the remaining monkeys, the roles of yellow and red were interchanged. The monkeys exhibited an aversion for the color paired with toxicosis on Day 5 (but not on later days). There is no obvious difference to a human between the flavor of the undiluted red coloring and that of the undiluted yellow coloring and the solutions contained only 0.08% coloring anyway. Thus the monkeys must have associated directly between the color of the solution and toxicosis.

Despite the modifications of the two-system approach of Garcia and Ervin to stimulus relevance which are demanded by the preceding findings, their approach still seems like a good beginning.

B. SITUATIONAL RELEVANCE

The Capaldi Effect, which was described in the introduction to this chapter, involves discriminative control of a running response by the reward outcome of the preceding trial. According to concurrent interference theory, it must depend on the relevance principle because a huge number of events are bound to occur during an intertrial interval of several hours. However, the relevance principle cannot possibly be stimulus relevance because many of these intervening events are bound to be potentially effective discriminative stimuli for a running response. These events include exteroceptive stimuli and movements of the animal which occur in the home cage between trials. Thus another type of relevance must be responsible for the Capaldi Effect; it may be called "situational relevance." Loosely speaking, situational relevance means that animals tend to associate directly from one situation to the next similar situation, just as they associate directly from flavors to toxicosis. More specifically, situational relevance means that the events which occur outside of the runway do not become associated with the events which occur inside the runway. If so, the outcome of the preceding trial can become an effective discriminative stimulus for the following trial because the intervening events produce little interference.

Unfortunately, there is no evidence for situational relevance as direct and unequivocal as the evidence for stimulus relevance shown in Fig. 1. However, in Chapter 3 of this volume, Capaldi shows that rats, trained on some trials in a black runway and on other trials in a white runway, associate events occurring, for instance, in the black runway with subsequent events occurring in the black runway rather than with subsequent events occurring in the white runway (see also Capaldi & Spivey, 1964). This suggests that association from one situation to another depends on the similarity of the situations. Therefore, if there is a delay between two elements of a learning paradigm and both elements occur in the same situation (such as a runway or a T-maze), concurrent interference will be reduced if the animal is removed from the experimental apparatus and placed in a discriminably different chamber during the delay. Thus the Capaldi Effect is presumed to depend on the removal of the rat from the runway during the intertrial interval; if the rat were to remain in the goal box during a very long intertrial interval, such events occurring in the goal box during the delay as the animal's movements would be expected to produce enough concurrent interference to prevent a discrimination based on the reward outcome of the preceding trial. Similarly, the findings of intertrial associations by Petrinovich and Bolles (1957) and by Pschirrer (described in Section I of this chapter) are presumed to depend on the removal of the rat from the apparatus during the intertrial interval.

B. T. Revusky (personal communication, 1971) has used removal of rats from the apparatus to obtain effective delay of reinforcement without the mediation of secondary reinforcers. Rats were taught a position habit in a modified T-maze. Just after a rat responded, it was removed from the apparatus without being rewarded and placed in its home cage regardless of whether its response was correct or incorrect. After a 1-minute delay the rat was returned to the start box. If its earlier response had been correct, the start box contained the day's ration of food. If its earlier response had been incorrect, the rat continued to receive choice trials with a 1-minute intertrial interval until it finally responded correctly and received its daily food ration in the start box. Of 15 rats, 14 performed better than chance during Days 11–20 of training. In a later experiment the delay was 5 minutes. Each of 8 rats performed better than chance during Days 41–50. Note that the rat's propioceptive set was disrupted by the handling which occurred after the response. Also, the rat remained in its home cage during the delay regardless of which response it had emitted. Thus differential secondary reinforcement cannot possibly account for these results, which had been predicted by B. T. Revusky on the basis of the theoretical ideas advanced in this chapter.

The term "situational relevance" is meaningful only in the context of the present attempt to explain all instances of association over long delays in terms of a single theory. It is not helpful in the analysis of the theoretical problems with which Capaldi is concerned, and he does not use "situational relevance" or any equivalent term. Capaldi does, however, use the principle of redintegration to explain how animals may associate directly from one trial to another. "Redintegration" means that if stimuli have occurred together in the past, presentation of some of these stimuli will tend to produce recall of the absent stimuli. Thus the stimuli in the start box, having previously become associated with the stimuli in the runway and in the goal box, produce recall of whether or not reward was obtained in the goal box on the preceding trial. In the present context, redintegration is an explanation of situational relevance. It must be emphasized, however, that while an explanation of situational relevance is desirable, concurrent interference theory is logically complete without any such explanation.

Finally, a subtle point must be made about the discriminative stimuli responsible for the Capaldi Effect. In experiments described in the introduction to this chapter, Pschirrer made chow pellets and milk differentially control the probability and/or direction of a running response. If the effective discriminative stimuli were the flavors of the two substances, then the proposed explanation of the Capaldi Effect contradicts the proposed explanation of the Garcia Effect. For the Garcia Effect has been hypothesized to depend, in part, on the low associative strength of flavors relative to events which emanate directly from the external environment. If so, flavors must be poor discriminative stimuli for running responses controlled by the receipt of food from the external environment.

The solution to this apparent contradiction depends on the assumption that the stimuli responsible for the Capaldi Effect are not flavors, but other stimuli correlated with reward. These other stimuli might be the activities of the animal upon receipt of the food reward; we have already discussed work by Garcia *et al.*

(1968) which suggests that such activities readily become associated with events which emanate directly from the external environment. If so, the stimulus differences between chow pellets and milk responsible for Pschirrer's results were probably the proprioceptive differences between seizing and chewing a pellet and licking milk. While the main basis for this conjecture is the proposed theory, it is noteworthy that in pilot work Pschirrer was not able to obtain a discrimination based on two flavors of liquid food, suggesting that flavors alone were not adequate discriminative stimuli for long-delay learning in his situation.

C. RESPONSE RELEVANCE

Seligman (1970) points out that in the first extensive experimentation involving instrumental learning in animals, R. L. Thorndike found it difficult to teach cats to scratch themselves or to lick themselves to escape from confinement even though these responses have a high operant level. Apparently these were not the types of responses which the cats could associate with escape. Although Seligman uses a different term, in the context of this chapter, it is useful to call such effects "response relevance."

Seligman argues that, in practice, throughout the history of experimentation in animal learning, the responses selected for investigation in different learning tasks were tacitly selected on the basis of response relevance because it was very difficult to train the animals when this was not done. Despite this, response relevance was not explicitly recognized as a central principle of learning theory. According to Revusky and Garcia (1970, pp. 48-49), response relevance is helpful for the interpretation of some work by Konorski (1967) and by Breland and Breland (1961). Furthermore, it is not impossible that the relevance of the feeding response to physiological aftereffects may occasionally be involved in the regulation of food intake although there is no evidence for this at present. Regrettably, the present understanding of the role of relevance is at best in an exceedingly primitive stage.

VI. Concurrent Interference in Flavor-Toxicosis Associations

The proposed theory is that long-delay learning is similar to other types of learning except that relevance has a central role. To confirm this, similarities between flavor-toxicosis associations and other types of learning will be described in this section.

Revusky and Garcia (1970, pp. 5-16) have summarized the evidence that learned aversions to flavors are affected by the various parameters of conditioning in much the same way as other behaviors with two exceptions: (1) Of course, the flavor aversions are susceptible to long delays of punishment and, (2) the learning is often so rapid that sometimes parametric effects are difficult to detect because the aversion is maximum at all levels of a parameter which are used. However, when this latter ceiling effect is not allowed to distort the results, the usual parametric effects emerge. The aversions become more pronounced as the severity of the contingent toxicosis is increased and with an increased number of training trials. They become less pronounced and extinguish when repeated consumption of the flavored substance is not punished. However, extinction of flavor aversions typically occurs more slowly than extinction of the usual locomotive and manipulative responses. The aversion also become less pronounced with an increased delay between ingestion of the flavored substance and toxicosis. Furthermore, there is also a reward effect which can determine the preference for flavors. If thiamine-deficient rats are fed a flavored solution and then are injected with thiamine, they exhibit an increased preference for the flavor (Campbell, 1969; Garcia, Ervin, Yorke, & Koelling, 1967; Zahorik & Maier, 1969). If hungry rats are intragastrically injected with food after they have consumed a nonnutritive substance, they develop an increased preference for that substance (Holman, 1969).

From the point of view of concurrent interference theory, however, these similarities between changes in the preferences for flavors and other behavioral changes are not the strongest evidence that the same learning mechanisms are responsible for both. It is always possible for different processes to produce similar parametric effects. More direct evidence that the same mechanism underlies both changes in preferences for flavors produced by toxicosis and the types of learning traditionally studied would be that the concurrent interference principle is applicable to both. Sections III, C and IV provide a reasonable case that concurrent interference occurs in traditional learning situations and Section V shows that long-delay learning is also explicable in terms of concurrent interference. However, Section V contains no direct evidence that concurrent interference actually occurs in long-delay learning situations; this omission will be remedied here.

A number of experiments will be reported in which two flavored solutions are consumed prior to toxicosis. For analytical purposes, one flavor will have the role of E-pre, the other will be E-pre-X, and the toxicosis will be E-post. The measure of the association of the E-pre flavor with toxicosis will be the subsequent preference for the E-pre substance; the lower the preference relative to appropriate controls, the greater the association. Variants of this basic procedure will be used to yield four types of experimental findings affirming that the concurrent interference principle applies to the association of flavors with toxicosis. (1) The more pronounced the flavor of E-pre-X, the greater the amount of concurrent interference produced (i.e., overshadowing will occur). (2) If the E-pre-X substance is consumed prior to E-pre instead of during the delay (the usual procedure in the experiments in this series), it will still produce concurrent interference. (3) If the E-pre-X substance becomes associated with

E-post prior to training, the amount of concurrent interference will be increased (i.e., blocking will occur). (4) If the E-pre-X substance is repeatedly consumed in the absence of E-post prior to training, so that it is less likely to become associated with E-post during training, the concurrent interference it produces will be reduced.

The order in which the experiments will be reported will correspond to the order in the preceding paragraph. Chronologically, the order in which each of these four topics was investigated happened to be exactly opposite to the order in which they are presented here.

A. OVERSHADOWING

Pavlov (1927) found that the more intense a CS is, the more rapidly it will become conditioned and the more overshadowing it will produce. Presumably, this also should be true for the intensity of a flavor stimulus so that the more intense a flavor, (1) the more rapidly it will become associated with toxicosis, and (2) the more interference it will produce in its role as E-pre-X. Garcia (1971) has confirmed the first of these implications. A test of the second implication will be reported below.

The gist of the experimental strategy was as follows: During conditioning, rats were first allowed to drink 0.2% saccharin solution (E-pre). Then different groups were allowed to drink different concentrations of vinegar solution (E-pre-X). Finally, all the rats were subjected to toxicosis (E-post) except for those in a nontoxicosis control group. Three days later, a series of preference tests for the saccharin flavor began. It was expected that the preference for saccharin solution among the experimental groups would increase with the concentration of vinegar consumed on the day of conditioning. This would show that the more intense the flavor of E-pre-X, the greater the interference with the aversion to the E-pre substance. After this first series of tests, a second conditioning trial was administered followed by a second series of tests.

The details of the experiment were as follows: Male Wistar rats, about 30 days old, were divided into five groups of 8–10. For 6 days, they were habituated to 1 hour daily of access to water. All training and testing, except for injections, occurred in the home cages.

On the first day of conditioning, all the rats were administered 2 ml of 0.2% saccharin solution in a water cup at their usual drinking time. Fifteen minutes after the saccharin solution was presented, all rats were administered 5 ml of a second solution. This second solution was unflavored tap water for one experimental group, while for the remaining three experimental groups, it contained 0.5%, 1.5%, or 4.5% by volume of Heinz brand cider vinegar. The nontoxicosis control group also received 4.5% vinegar solution. One hour after the vinegar solution was administered, all the experimental rats were injected intraperitoneally with 20 ml/kg of 0.15 M lithium chloride solution in order to produce toxicosis. The nontoxicosis control rats were not injected at all.

During the two days following the first day of conditioning, the rats were permitted their usual hour of free access to tap water. On each of the following 3 days, there was a 1-hour test of saccharin preference. An unusual procedure was used. Usually, the rats would be given a choice between saccharin solution and distilled water. In this experiment, however, a 1.25% (w/v) solution of Sanka instant decaffeinated coffee was substituted for the water because there was a chance that the conditioning procedure might have produced an aversion to water in the experimental group which received water between the saccharin solution and toxicosis on the day of training. Preference for saccharin solution was defined as the weight of saccharin solution consumed divided by the total weight of fluid consumed.

On the day following the first series of three tests, a second conditioning trial was administered followed by 7 days more of testing.

Figure 2 shows the mean preference for saccharin solution for each group on each test day. The lower the preference for saccharin solution, the greater is the learned aversion to the flavor of saccharin.

Figure 2 indicates the aversion to the saccharin E-pre flavor became less pronounced as a function of the concentration of vinegar in the E-pre-X solution. Statistical analysis confirmed the reliability of this trend both for Test Days 1-3, prior to the second training trial, and Test Days 4-10. The datum used was the mean of the preferences obtained over the test days to be analyzed and the test was the F test for linear trend as a function of vinegar concentration because it is sensitive to monotonic effects. For Test Days 1-3, $F = 7.18$, 1/34 df, $p < .01$ for the expected direction of the trend; for Test Days 4-10, $F = 3.83$, 1/32 df, $p < .05$. A test between the no-toxicosis control group and the 4.5% vinegar experimental group yielded $t = 2.44$, 15 df, $p < .025$ for Test Days 1-3 and insignificant results for Test Days 4-10 ($.05 < p < .10$).

The overshadowing shown in Fig. 2 was not as complete as much of the overshadowing reported by Pavlov (1927). While it is possible that this is a genuine difference between flavor-toxicosis associations and the types of associations investigated by Pavlov, it also is possible that this difference is due to a methodological inconvenience of using flavored substances as E-pre-X's. If the flavor of an E-pre-X substance is made very intense, the rat will not drink it; the animal has no such option in the case of a telereceptive stimulus. Conceivably, if the rats were somehow to be forced to drink undiluted vinegar, more pronounced overshadowing might be obtained.

It might seem that the data in Fig. 2 can be explained by an aftertaste hypothesis as well as by overshadowing. That is, the aftertaste of saccharin at the time of toxicosis might be expected to have a role in the learning of the aversion; the stronger vinegar solutions might be expected to produce greater interference

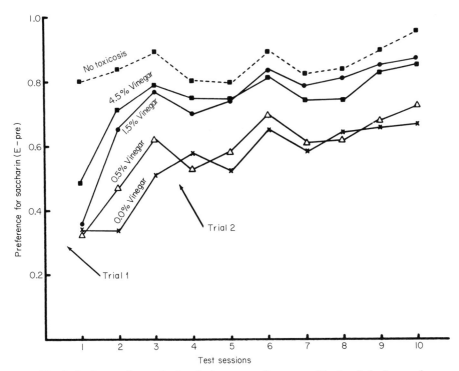

Fig. 2. Preference for saccharin solution among five groups. The top dashed curve shows the control level, when saccharin solution had never been followed by toxicosis. The four bottom solid curves show the preferences when different concentrations of vinegar solution intervened between saccharin consumption and toxicosis. Arrows show when training trials were administered.

with this aftertaste. However, in Section VI, D of this chapter, it will be shown that the type of interference demonstrated in Fig. 2 can be nearly eliminated if the rats are made familiar with the flavor of vinegar prior to the experiment so that its associative strength is reduced. Familiarization with vinegar solution should not be expected to interfere with its ability to destroy the aftertaste of saccharin. Furthermore, Section VI, B will show that interference can occur if the E-pre-X substance is consumed prior to the E-pre substance. If so, and if the interference is to be attributed to destruction of the aftertaste of the E-pre substance, then the destruction must have been caused by the aftertaste of the E-pre-X substance, and this is unreasonable. In general, while the aftertaste hypothesis may seem reasonable to those weaned on temporal contiguity theory, there is ample evidence that delayed flavor–toxicosis associations can occur in the absence of aftertastes at the time of toxicosis (Revusky & Garcia, 1970, pp. 17–18).

Attention theory fares even more poorly than the aftertaste hypothesis as an explanation of the data in Fig. 2. The concentration of the vinegar solution could not have affected the attention paid by the rats to the saccharin flavor because the saccharin solution was entirely consumed before the vinegar solution was presented.

B. INTERFERENCE BY E-PRE-X'S WHICH PRECEDE E-PRE

The preceding section showed that an E-pre-X which intervenes between E-pre and E-post can produce concurrent interference. The purpose of the two experiments reported in this section was to show that an E-pre-X can precede E-pre and still produce interference. The designs of both experiments had important elements in common. These will be discussed first, and only later will the procedures for the individual experiments be described.

The gist of the training procedure used in both experiments is outlined in Table II. There were two experimental groups and two control groups. First, 2 ml of the indicated E-pre-X solution (saccharin or unflavored tap water) were administered to the rats, which were thirsty enough to insure that all liquid solutions would be consumed with alacrity. Fifteen minutes later, 5 ml of the indicated E-pre solution (coffee or vinegar flavored) were made available. One hour later, toxicosis was induced by intraperitoneal injection of lithium chloride solution.

Table II

Design of Experiment to Demonstrate That E-pre-X Can Precede
E-pre and Still Produce Interference

Type of treatment	E-pre-X	E-pre
Experimental	Saccharin	Coffee
Experimental	Saccharin	Vinegar
Control	Unflavored water	Coffee
Control	Unflavored water	Vinegar

The tests of learning occurred some days after training. The rats were given a choice between the coffee and vinegar solutions for 1 hour while they were thirsty. The preference for E-pre was defined as the amount consumed of the E-pre solution divided by the total weight of solution consumed. With this method, the mean preference among both experimental groups or both control groups will approximate 0.5 if there is no aversion and will become progressively lower as the aversion to the E-pre substance becomes more marked (Rozin, 1969).

The experimental hypothesis was that the preference for the E-pre flavor would be lower among the control groups than among the experimental groups because saccharin solution would produce more interference than unflavored water. To increase the likelihood that this hypothesis would be confirmed, the rats were made familiar with the coffee and vinegar E-pre solutions prior to training; each solution was substituted for water in the home cages for 24 hours. It is well established that such familiarization with a flavor reduces the strength of a later association of that flavor with toxicosis (Revusky & Garcia, 1970, pp. 32–38; this chapter, Section VI, D). Since the saccharin solution was novel at the time of training, this familiarization was designed to insure that the saccharin E-pre-X solution would tend to become associated with toxicosis more readily than either E-pre solution and thus produce a large amount of concurrent interference. The control E-pre-X substance, unflavored water was designed to produce little interference because of its familiarity and its weak flavor.

The use of unflavored water as E-pre-X for the controls may seem like an unnecessary complexity. The purpose of the experiment was simply to show that E-pre-X can precede E-pre and still produce interference, so that it might seem more straightforward simply to compare the presence of E-pre-X with its absence. Unfortunately, this simpler procedure would result in less consumption of fluid by the control rats than by the experimental rats prior to toxicosis. Although this minor difference was ignored in experimentation to be reported in later sections, in the present case it could produce a serious artifact. Conceivably, a greater thirst level might increase the severity of the lithium chloride toxicosis and thus result in a more pronounced aversion among the controls regardless of whether the saccharin E-pre-X flavor produced concurrent interference.

1. Experiment 1

There were four groups of 8 to 10 male Wistar rats about 40 days old. Prior to the experiment proper, each rat received 24 hours of free access to 1.25% (w/v) of a solution of Sanka instant decaffeinated coffee as its only source of fluid. For the following 24 hours, 3% (w/v) of Heinz cider vinegar solution was substituted. For the following two days, the rats received 1 hour of water per day.

For the first 9 days of the experiment, the following 3-day cycle was repeated three times. (1) The first day was a training day. Two milliliters of the E-pre-X substance (0.2% saccharin solution or tap water) were administered in a water cup at the usual drinking time. Fifteen minutes afterward, 5 ml of the coffee or vinegar E-pre solution were administered. One hour later, toxicosis was induced (20 ml/kg of 0.15 M lithium chloride solution ip). (2) On the second day, 3 hours of free access to tap water was permitted. (3) On the third day, the rats were given a 1-hour choice test between the coffee and vinegar solutions.

After the third repetition of this 3-day sequence, 1-hour coffee–vinegar tests were administered every second day and no water on the intervening day. There were 25 tests including the first three.

Figure 3 shows the experimental results. Note that the first test occurred after the first conditioning trial so that the means were already below the chance level of 0.5. The generally higher level of the experimental scores agrees with the

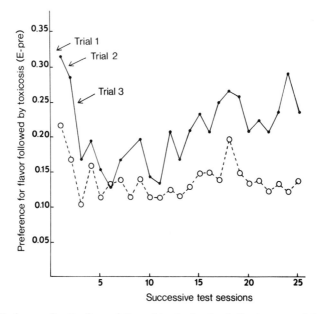

Fig. 3. Preference for the flavor followed by toxicosis relative to a second flavour when the two flavors were coffee and vinegar. The arrows refer to the occurrence of training trials between two test trials. The experimental rats (●–●) consumed 0.2% saccharin solution prior to the E-pre substance during training; the control rats (○–○) consumed tap water.

hypothesis that the experimental treatment of administering novel saccharin solution prior to E-pre would interfere with the aversion to the coffee or vinegar E-pre flavor. A 2 x 2 F test showed that during the first test, the experimental scores were significantly higher than the control scores ($F = 3.60$, 1/32 df, $p < .05$, one tail). During the second test, the scores no longer were reliably different ($F = 2.82$, $.05 < p < .06$). During the next few tests, both treatments yielded substantial aversions to the E-pre substance so that a floor effect prevented detection of any differences between them.

There was remarkably little extinction of the aversions to the E-pre substance over the long series of test sessions. Visual inspection of Fig. 3 gives the impression that the experimental treatment resulted in some extinction while the control treatment resulted in none. But inspection of the raw data from the twenty-fifth session shows that more than half of the experimental rats exhibited aversions as pronounced as those obtained on the fourth session; only a minority of the experimental rats were responsible for the rise in the curve. Thus the rats received their last training trial about 4 weeks after they were weaned and grew nearly to maturity without any substantial changes in their preferences although their entire water intake was in the form of coffee and vinegar solution. Such slow extinction was not apparent in the results of Fig. 2. This difference can be attributed to any one of a number of differences between the two procedures.

2. Experiment 2

The results shown in Fig. 3 did not demonstrate the production of interference by an E-pre-X which occurs prior to E-pre as powerfully as had been desired. To obtain a more pronounced effect in Experiment 2, a number of changes were made in the experimental procedure. The most important of these was an increase in the concentration of the saccharin E-pre-X solution from 0.2% to 2.0%. Presumably, this increase in concentration would increase the intensity of the flavor which, in turn, would increase the associative strength of the saccharin solution, and hence increase the amount of concurrent interference.

Each of the four groups contained 15 rats of the same type and age as those used in Experiment 1. Prior to the experiment proper, 1.25% coffee solution was substituted for water for 24 hours and then 3.0% vinegar solution was substituted for another 24 hours. After this, there were two 1-hour choice tests between the coffee and vinegar solutions on two successive days; the first of these tests followed 2 days of water deprivation. The four groups were equated for preexperimental preferences on the basis of the second of these tests.

The experiment proper began on the day following the second of these preliminary choice tests. On Day 1, the rats were trained by a procedure identical to that of Experiment 1 (Table II) except that the concentration of the saccharin solution was 2.0% and the toxicosis was made slightly more severe. That is, 2 ml of the E-pre-X substance (saccharin solution or unflavored water) were administered 15 minutes before 5 ml of the E-pre solution (coffee or vinegar). Toxicosis was induced an hour later by injection of 25 ml/kg of 0.15 M lithium chloride. On Days 2 and 3 water was made available for 1 hour per day. On Days 4, 5, and 6 there were 1-hour choice tests between coffee and vinegar solution; no other source of water was available on those days.

On Day 7 of the experiment, a second training trial was administered. On Day 8 water was made available for 1 hour. On Days 9–13 inclusive, there were daily 1-hour coffee–vinegar choice tests.

Figure 4 shows the results. After the second training trial, each test day yielded a significant treatment effect ($p < .01$, one tail) according to a 2 x 2 F

test. Thus the saccharin flavor overshadowed the coffee or vinegar flavor even though the saccharin solution was consumed prior to the coffee or vinegar solution.

The magnitude of the interference shown in Fig. 4 encourages the conjecture that conditions may exist in which the interference obtained in flavor-toxicosis associations might be as complete as that obtained by Pavlov (1927) and by

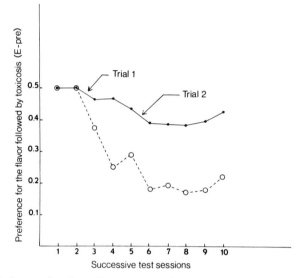

Fig. 4. Preference for the coffee or vinegar flavor followed by toxicosis. The arrows refer to the occurrence of a training trial between two test trials. For the experimental rats (●—●), 2.0% saccharin solution was consumed prior to the E-pre substance during training; the control rats (○—○) consumed tap water.

Kamin (1969) in more conventional experiments. Indeed, when the results of the three tests occurring between the first and the second training trials were combined, the experimental rats did not exhibit a significant aversion to the E-pre flavor. (Statistical reliability was assessed by expressing the preferences for each experimental group in terms of the preference for coffee and comparing the two groups by means of a t-test: $t = 1.14$, 28 df, $p > .10$, one tail.) However, the experimental groups did exhibit an aversion to the E-pre flavor during the five test sessions following the second training trial ($t = 2.13$, 28 df, $p < .025$, one tail).

There was one noteworthy unexpected result. The third test session followed the first training trial and there were no additional training trials until after the fifth test session. Thus there was no reason to expect a drop in the preference for E-pre between the third and the fourth test session. Nevertheless, 29 of 30 control rats did exhibit such a drop. This result has a probability of below 1 in 17,000,000 according to the binomial theorem which is low enough to justify the risk of discussing an unreplicated and unexpected result. The drop in preference shows that one session of exposure to the E-pre flavor without any experimentally produced toxicosis may result in an increase of the aversion to E-pre. This is reminiscent of an effect reported by Solomon and Wynne (1953) for dogs which learned to jump in response to a signal which terminated in shock if they failed to respond within 10 seconds. After the latencies were reliably below 10 seconds so that shock was avoided on all trials, there still was a further reduction in latency indicating that the dogs learned to jump faster to the signal even in the absence of shock. Similarly, the control rats of Fig. 4 learned a more pronounced aversion to E-pre in the absence of a second induction of toxicosis. Here is still another similarity between flavor–toxicosis associations and other types of learning.

C. BLOCKING

Kamin's demonstration (1969) of blocking and its explanation by the concurrent interference principle has already been discussed. Blocking occurs when E-pre-X becomes associated with E-post prior to training; since the association of E-pre-X with E-post is already in existence at the beginning of training, it tends to prevent the reference association from developing. R. G. Van Houten (personal communication) suggested that a similar effect might occur if an E-pre-X flavor were associated with toxicosis prior to training; during training, it too might tend to prevent the E-pre flavor from becoming associated with toxicosis. This expectation has been confirmed by two experiments which will be reported below.

In both experiments saccharin solution (E-pre) was consumed and then either coffee or vinegar solution (E-pre-X); afterward, toxicosis (E-post) was induced. In the blocking treatment the coffee or vinegar E-pre-X had previously been paired with toxicosis; in the no-blocking treatment, the same flavor (E-pre-X) had not previously been paired with toxicosis. The test of conditioning was preference for saccharin solution relative to water. If Kamin's blocking effect is applicable to flavor–toxicosis associations, then the blocking treatment must produce more interference with the reference association than the no-blocking treatment and, hence, result in a higher preference for the saccharin solution.

The experimental design was more complex than the above outline may suggest because it was desirable that the no-blocking control treatment be equivalent to the blocking treatment in the following ways: (1) Amount of exposure to toxicosis during pretraining. Conceivably, prior exposure to toxicosis, even if the toxicosis is not paired with E-pre-X, might change the subsequent capacity of toxicosis to produce an aversion to a flavor. (2)

Experience with the E-pre-X flavor during pretraining. The familiarity of a flavor is an important determinant of its associative strength relative to toxicosis (Revusky & Garcia, 1970, pp. 32–38; Section VI, D of this chapter) so that it was desirable that E-pre-X be equally familiar under both treatments.

Table III shows how these desiderata were met by the experimental design. All rats drank both coffee solution and vinegar solution during pretraining and one of these solutions was paired with toxicosis. The design was 2 x 2 factorial in the first experiment. The first factor was whether the E-pre-X flavor was coffee or vinegar. The second factor was whether or not the E-pre-X flavor had been paired with toxicosis during pretraining. Of course, it is the second factor which is of theoretical interest.

Table III

Design of Experiment to Show That Blocking Occurs in Flavor–Toxicosis Learning

Type of treatment	Pretraining procedure	Training procedure
Blocking	Coffee solution (E-pre-X) is paired with toxicosis, while vinegar solution is consumed without contingent toxicosis	Coffee (E-pre-X) is consumed between saccharin and toxicosis
Blocking	Vinegar solution (E-pre-X) is paired with toxicosis, while coffee solution is consumed without contingent toxicosis	Vinegar (E-pre-X) is consumed between saccharin and toxicosis
No-blocking (control)	Vinegar solution is paired with toxicosis, while coffee solution (E-pre-X) is consumed without contingent toxicosis	Coffee (E-pre-X) is consumed between saccharin and toxicosis
No-blocking (control)	Coffee solution is paired with toxicosis, while vinegar solution (E-pre-X) is consumed without contingent toxicosis	Vinegar (E-pre-X) is consumed between saccharin and toxicosis

1. Experiment 1

The details of the procedure were as follows: Male Wistar rats, 35–40 days old, were maintained on a water deprivation schedule and divided into four groups of 8–10. Pretraining was given during Days 1–7. On Day 1 toxicosis (20 ml/kg of 0.15 M lithium chloride ip) was induced for all rats 30 minutes after the end of a 30-minute period of access to either 1.25% coffee or 3.0% vinegar solution. On Days 2 and 3 the coffee or vinegar solution, whichever had not been administered on Day 1, was made available for 30 minutes per day; toxicosis was not induced. On Day 4 the rats received 4 hours of free access to tap water. They were not watered on Day 5. On Day 6 coffee or vinegar solution was paired with toxicosis in the same way as on Day 1. On Day 7 tap water was made available for 4 hours.

Training occurred on Day 9, 2 days after the rats had last received water. Each rat received 2 ml of 0.2% saccharin solution (E-pre) followed, 15 minutes later, by 5 ml of coffee or vinegar solution (E-pre-X). After another hour toxicosis (E-post) was induced. All the rats consumed their 2 ml of saccharin solution within 15 minutes and hence prior to the administration of the coffee or vinegar E-pre-X solutions. About half of the rats subjected to the blocking treatment did not complete consumption of their 5 ml of coffee or vinegar E-pre-X solution during the hour prior to toxicosis because, of course, they had a learned aversion to it; for these animals, the fluid was removed just before injection. The rats subjected to the no-blocking treatment drank their E-pre-X solution with alacrity.

On Day 10, tap water was made available for 4 hours; the test of conditioning was given on Day 11. The rats were given 23 hours of free access to 0.2% saccharin solution and to distilled water. Saccharin preferences (saccharin intake as a proportion of total fluid intake) were calculated separately for hours 0–1, 1–7, and 7–23.

Figure 5 shows the experimental results. During each of the three parts of the test, the blocking treatment resulted in the expected higher preference for the

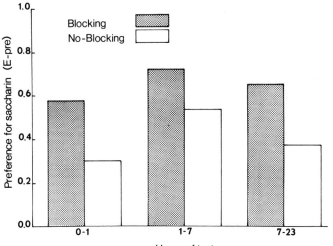

Fig. 5. Comparison of the preferences for saccharin solution produced by two treatments. In both treatments the rats first consumed saccharin solution, then coffee or vinegar solution, and then were subjected to toxicosis. In the blocking treatment, the intervening coffee or vinegar solution had previously been associated with toxicosis. In the no-blocking treatment, the intervening solution had not previously been associated with toxicosis.

saccharin E-pre solution than the no blocking treatment. Statistical assessment was based on the mean of the three preferences obtained from the separate parts of the test. An F test was used to factor out whether coffee or vinegar had the role of E-pre-X and yielded $p < .005$, one tail ($F = 7.98$, 1/34 df) for the treatment effect. Thus, as might be expected on the basis of the concurrent interference principle, the Kamin blocking effect applies to flavor–toxicosis associations.

2. Experiment 2

Experiment 2 was more elaborate than Experiment 1 because, as is shown in Table IV, two treatments were used in addition to the blocking and no-blocking treatments.

Table IV

The Four Treatments Used in Experiment 2 about Blocking

Type of treatment	Training procedure
No E-pre-X	Consumption of saccharin solution is followed by toxicosis and nothing is consumed during the delay
No-blocking	Coffee or vinegar solution, not previously paired with toxicosis, is consumed during the interval between saccharin consumption and toxicosis
Blocking	Coffee or vinegar solution, previously paired with toxicosis, is consumed during the interval between saccharin consumption and toxicosis
No E-post	No toxicosis is induced after saccharin consumption, but the E-pre-X solution used in the blocking treatment is consumed

1. In the no E-pre-X treatment, when saccharin solution was paired with toxicosis, no intervening substance was consumed. This treatment was expected to produce a more pronounced aversion to the saccharin E-pre solution than the no-blocking treatment because even though the E-pre-X flavor of the no-blocking treatment was not paired with toxicosis prior to training, it still might be expected to produce some interference.

2. The no E-post treatment was similar to the blocking treatment except that toxicosis was not induced on the day of training; that is, the rat first drank saccharin solution (E-pre) and then the E-pre-X solution which had been paired with toxicosis during pretraining. The purpose of this treatment was to assess the magnitude of the blocking effect; for if blocking was complete, the blocking treatment and the no E-post treatment would result in similar preferences for the saccharin E-pre solution.

Experiment 1 contained two procedural deficiencies which conceivably might have produced an invalid inflation of the difference between the results of the blocking and the no-blocking treatments. These were corrected in Experiment 2 as follows:

1. During the pretraining phase of Experiment 1, the rats had two 30-minute exposures to the E-pre-X solution paired with toxicosis and two 30-minute exposures to the E-pre-X solution not paired with toxicosis. However, the rats drank relatively little during the second exposure to the flavored solution paired with toxicosis because they had already developed an aversion to it. This may mean that the flavor paired with toxicosis was less familiar to the rats than the other flavor. Since a less familiar flavor produces more interference than a more familiar flavor, the magnitude of the difference between the effects of the blocking and the no-blocking treatments in Experiment 1 may have been increased due to this difference in familiarity. During the pretraining phase of Experiment 2, the amount consumed of the coffee or vinegar solution not paired with toxicosis was fixed at 5 ml for each exposure; thus the rats actually consumed less of this E-pre-X solution than of the E-pre-X solution which was paired with toxicosis.

2. Since the blocking treatment of Experiment 1 consisted of training the rats with an E-pre-X substance to which they had previously developed an aversion, rats subjected to that treatment drank the E-pre-X solution more slowly during training than rats subjected to the no-blocking treatment. Thus the blocking treatment produced a longer exposure to the E-pre-X substance. Conceivably, the longer the exposure to an E-pre-X substance, the greater the strength of its association with toxicosis, and the more concurrent interference it produces. A blocking effect attributable to such a process would differ from Kamin's blocking effect because Kamin (1969) used a fixed duration of E-pre-X presentation. In Experiment 2 the difference in exposure times to E-pre-X under the two treatments was reduced; the cup containing the E-pre-X solution was removed 15 minutes after it had been presented even if the 5-ml ration had not been completely consumed. On the average, the E-pre-X substance still was available to the blocking rats for a longer period than it was available to the no-blocking rats, but hopefully this was balanced by the fact that the blocking rats actually spent less time in contact with the E-pre-X solution because they had an aversion to it.

The procedure for Experiment 2 was as follows: Male Wistar rats, about 45 days old, had been maintained on a water deprivation schedule prior to the experiment. The flavored solutions and the type of toxicosis used were the same as in Experiment 1.

Pretraining occurred during Days 1-9. On Day 1 half of the rats were permitted to drink coffee solution for 30 minutes and the other half were permitted to drink vinegar solution. Fifteen minutes after the bottles were removed, toxicosis was induced in all rats. On Day 2 the rats were allowed 4 hours of access to tap water. On Days 3 and 4 the rats were given 5 ml per day of the solution (coffee or vinegar) which had not been paired with toxicosis on Day 1. On Day 5 the rats were given 4 hours of access to tap water and no additional fluid was administered until Day 7; then the procedure of Day 1 was repeated, so that either coffee or vinegar solution was paired with toxicosis. On Days 8 and 9 the rats received 4 hours of free access to tap water.

Training was administered on Day 11 after 2 days of water deprivation. Eleven rats were subjected to the no E-post treatment and 15 or 16 were subjected to each of the other three treatments; the roles of coffee and vinegar were appropriately counterbalanced. All rats first received 3 ml of saccharin (E-pre). Except for those subjected to the no E-pre-X treatment, the rats received 5 ml of the coffee or vinegar E-pre-X solution as appropriate 30 minutes later. The coffee or vinegar was removed after another 15 minutes. After still another 15 minutes, toxicosis (E-post) was induced except, of course, for the rats subjected to the no E-post treatment. On Day 12 the rats were permitted 4 hours of access to tap water.

Testing was done on Days 13–18, when the rats were given a choice between saccharin solution and distilled water for 1 hour per day.

Figure 6 shows the mean saccharin preferences for each treatment on each test day. The results were exactly in the predicted direction. The four treatments were not compared in a single overall analysis of variance because the difference

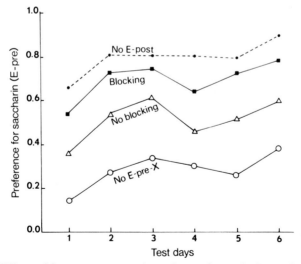

Fig. 6. Effects of four treatments on the preference for saccharin solution. See Table IV for a description of these treatments.

between the no E-pre-X treatment and the no E-post treatment was bound to produce significance, but this is of no relevance to the experimental hypotheses. Instead, two treatments were compared at a time by means of F tests which factored out the different roles of coffee and vinegar in pretraining; the datum used was the mean of the six test scores obtained for each rat. By this criterion, each treatment produced preferences which differed significantly from the preferences produced by every other treatment. The differences between treatments will be discussed in more detail below.

The difference between the no-blocking treatment and the no E-pre-X treatment was significant at the one-tailed .005 significance level ($F = 8.23$, $1/28\ df$). Figure 6 shows that the magnitude of this difference was quite large; this indicates that the E-pre-X which had not previously been paired with toxicosis produced a great deal of interference. This interference probably occurred because the rats were given only limited exposure to this E-pre-X solution prior to training; that is, only 5 ml on each of two occasions. It is known that more extensive familiarization with an E-pre-X flavor results in a more marked diminution of the amount of interference it produces (Revusky & Garcia, 1970, pp. 32–38).

Despite the experimental precautions which loaded the experiment against detection of a difference between the blocking treatment and the no-blocking treatment, there was a significant difference between the preferences they produced at the .025 level ($F = 4.29$, $1/27\ df$). Here again is support for the hypothesis that if an E-pre-X substance has been paired with E-post prior to training, it interferes with the reference association more than an E-pre-X substance not previously paired with E-post.

Figure 6 shows that the rats subjected to the blocking treatment exhibited preferences closer to those exhibited by rats not subjected to toxicosis during training (no E-post treatment) than to those exhibited by rats which were not subjected to any interference during training (no E-pre-X treatment). This shows that the magnitude of the blocking effect was very great. However, blocking was not complete because the blocking treatment yielded preferences significantly lower than those produced by the no E-post treatment ($F = 4.28$, $1/22\ df$, $p < .05$, one tail).

Finally, Figs. 5 and 6 are examples of blocking which cannot readily be explained in terms of lack of attention to the blocked E-pre. The rats subjected to the blocking treatment, to the no-blocking treatment, and to the no E-pre-X treatment all consumed the saccharin E-pre solution under exactly the same conditions; that is, the E-pre-X solution was not presented until after the saccharin solution was entirely consumed. If attention theory is to explain the differences among the results of these treatments, it must postulate that the E-pre-X flavor retroactively affected the amount of attention paid to the E-pre flavor. This seems farfetched in comparison to concurrent interference theory.

D. Reduction of Interference by Latent Inhibition of E-pre-X

It is of obvious evolutionary importance that animals avoid food and drink which produce sickness. It is also important that they do not avoid harmless substances which happen by chance to be consumed prior to sickness. If as omnivorous an animal as the rat were to avoid everything which it had consumed prior to a sickness produced by a slow-acting poison, it might well starve to death. The blocking effect described in the preceding section is a very efficient means of preventing such indiscriminate association. If there is a substance which has produced sickness in the past among the substances consumed prior to another instance of a similar sickness, the chances are that the other substances are safe to consume. The concurrent interference principle ensures that there will be less of an aversion to these other substances than if the animal had not consumed a substance which had produced sickness in the past.

There is a second, probably more important mechanism allowing highly adaptive selective association of flavors with later sickness. Obviously, if there is a novel substance among those consumed prior to sickness and the other substances have been repeatedly consumed in the past without producing sickness, then the novel substance is probably responsible for the sickness. It will be shown below that a novel substance will tend to become more readily associated with toxicosis than a familiar substance. Hence it will tend to prevent the familiar substance from becoming associated with the toxicosis.

1. Latent Inhibition of Flavor–Toxicosis Learning

I have already referred to the latent inhibition effect: repeated presentation of an E-pre in the absence of E-post decreases the associative strength of E-pre relative to E-post. In the present context, this implies that a flavor aversion produced by ingestion-contingent toxicosis will be less pronounced if the flavor is familiar than if it is novel.

There is much indirect evidence that latent inhibition of flavors occurs; for aversions have often been obtained under conditions in which the rats consumed familiar substances during the delay of punishment interval. For example, in the experiments whose results are reported in Figs. 2–6, the rats were maintained on a limited water supply with dry chow continuously available even during the delay of punishment interval. The rats almost certainly ate chow during the delay because rats with restricted access to water tend to eat shortly after drinking. If latent inhibition had not been operative, the toxicosis would have become associated with the flavor of the chow; and this association would have interfered with the association between the E-pre solution and the toxicosis. Furthermore, other relevant events, such as internal stimuli and smells, are likely to have occurred during the delay. Without the occurrence of latent inhibition, these presumably familiar, relevant stimuli would have produced interference.

Farley, McLaurin, Scarborough, and Rawlings (1964), Garcia and Koelling (1967), McLaurin, Farley, and Scarborough (1963), Nachman (1970), Revusky and Bedarf (1967), and Wittlin and Brookshire (1968) have all supplied direct evidence that aversions to familiar flavors produced by ingestion-contigent toxicosis are less pronounced than aversions to novel flavors. The experiment of McLaurin *et al.* (1963) will be described: The concern was with the changes in the preference for saccharin solution produced by ingestion-contingent X-irradiation. In each of five groups, mature rats drank 0.1% saccharin solution for 30 minutes following 24 hours without water. The rats in Group No-Hab, which had not been habituated to the saccharin solution prior to training, were subjected to 102 r of irradiation shortly after the saccharin drinking period. Since there was no latent inhibition effect to attenuate the aversion to the saccharin, this group was expected to show a maximal aversion. The rats in Group Sham-Irr were not irradiated after drinking the saccharin solution and thus were not expected to exhibit any aversion to saccharin. The remaining three groups received 6 days of continuous access to saccharin solution prior to conditioning. These groups differed in the amount of time intervening between the latent inhibition procedure and the day of conditioning. For Group Hab-0, the 6 days of latent inhibition terminated on the day before irradiation. Group Hab-3 received 3 days of continuous access to unflavored water between the latent inhibition to the saccharin solution and conditioning. Similarly, Group Hab-6 received 6 days of access to water. Table V summarizes the procedures for all groups except for the sham irradiated controls.

Shortly after irradiation, all groups were given a choice between saccharin solution and distilled water for 7 days. Figure 7 shows the preferences for saccharin among the five groups during the 7-day test. The latent inhibition effect was obtained; each of the three groups which were made familiar with the saccharin solution prior to conditioning, exhibited a less pronounced aversion to saccharin than did Group No-Hab, which was not exposed to saccharin solution prior to training. However, all the latently inhibited groups exhibited some aversion to saccharin relative to the sham irradiated controls (Group Sham-Irr). Thus latent inhibition did not completely prevent association of the saccharin flavor with toxicosis. The differences among Groups Hab-0, Hab-3, and Hab-6, as shown in Fig. 7, seem to indicate that the latent inhibition effect became less pronounced as a function of time between the latent inhibition and conditioning.

The proposed theory is logically complete without any explanation of latent inhibition itself. It will be treated as a parameter affecting associative strength without any commitment to further explanation. This is the same way in which psychophysical intensity was dealt with earlier. However, it is possible that as concurrent interference theory develops, it will become capable of explaining latent inhibition. For instance, in the course of familiarization, the flavored

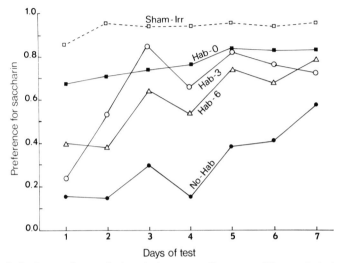

Fig. 7. Preference for saccharin solution among five groups. The top dashed curve is for rats which had never been subjected to toxicosis after drinking saccharin solution. See Table V for a description of the other treatments. (Redrawn from McLaurin *et al.,* 1963.)

Table V

Pretraining Treatments Used by McLaurin, Farley and Scarborough (1963) for the Irradiated Groups

Groups	Pretraining procedure until day prior to irradiation
No-Hab	No preexposure to saccharin solution
Hab-0	Saccharin solution for 6 days
Hab-3	Saccharin solution for 6 days followed by tap water for 3 days
Hab-6	Saccharin solution for 6 days followed by tap water for 6 days

substance may become associated with some beneficial E-post-X, such as alleviation of thirst. Such an E-post-X could also occur concurrently with toxicosis during training and thereby interfere with the association with E-post. If so, latent inhibition is closely related to blocking.

Latent inhibition may seem easily explicable by a straightforward attention theory on the basis that familiarization reduces the attention the animal pays to E-pre. This explanation is not tenable for the latent inhibition of flavors because it has frequently been observed that rats developed increased preferences for familiar flavors with repeated exposure to them (Revusky & Bedarf, 1967; Revusky & Garcia, 1970, pp. 63–64; Wittlin & Brookshire, 1968). It is hard to conceive how preference for a flavor can be increased without the animal's attending to it. Thus the reduction in associative strength produced by latent inhibition cannot be the result of lack of attention.

2. Interaction of Latent Inhibition with Concurrent Interference

If latent inhibition reduces the associative strength of a flavor relative to toxicosis, according to the concurrent interference principle it must also reduce the interference produced by that flavor when it has the role of E-pre-X. Revusky, Lavin, and Pschirrer confirmed this expectation in experimental work described by Revusky and Garcia (1970, pp. 34–38).

Using 4-month-old, male Wistar rats, they performed two closely related experiments which differed only in the nature of the E-pre substance: 10% (w/v) sucrose solution or 0.2% (w/v) saccharin solution. There were two E-pre-X solutions in each experiment: 1.0% (w/v) coffee solution and 3% (w/v) vinegar solution. For all rats, one of these solutions was the only source of water for 8 days and they were allowed 1 hour per day to drink. Training occurred 2 days after the final day of familiarization; during these 2 days the rats had no access to any fluid. Four treatments were administered:

1. No E-pre-X. Four ml of the saccharin or sucrose E-pre solution were administered in a water cup. Toxicosis was induced 100 minutes after E-pre was presented by ip injection of 20 ml/kg of 0.15 M lithium chloride solution.

2. Familiar E-pre-X. This treatment differed from the above in that 4 ml of the familiar E-pre-X was administered 50 minutes after the administration of E-pre and 50 minutes prior to the injection. The roles of the coffee and vinegar flavors were appropriately counterbalanced.

3. Novel E-pre-X. This differed from the previous treatment in that the novel E-pre-X was used instead of the familiar E-pre-X.

4. No E-post. Toxicosis was not administered. In other respects, half the rats were treated as the familiar E-pre-X rats and half were treated as the novel E-pre-X rats.

On the day after training, the rats were given free access to tap water for 4 hours. On the following day, they were given a 23-hour choice test between the saccharin or the sucrose E-pre solution and distilled water. Preferences were computed separately for hours 0–1, 1–3, and 3–23.

The results will be reported here only insofar as they are relevant to the concurrent interference approach because they have already been reported in more detail elsewhere (Revusky & Garcia, 1970, pp. 34–38). Table VI shows the preference for E-pre produced by each treatment when E-pre was saccharin and when it was sucrose. Each datum is a mean of scores from 11 or 12 rats and each score was a mean of the three preference scores obtained for different test intervals.

There was no overall F test of the data shown in Table VI because a

Table VI

Results of Experiment by Revusky, Lavin, and Pschirrer[a]

Type of treatment	Preferences for different types of E-pre	
	Saccharin	Sucrose
No E-pre-X	0.33	0.53
Familiar E-pre-X	0.32	0.59
Novel E-pre-X	0.49	0.70
No E-post	0.69	0.74

[a] Reported by Revusky and Garcia (1970).

significant treatment effect was bound to be obtained due to the already well-documented difference between the no E-pre-X treatment and the no E-post treatment. Instead, two treatments at a time were compared by means of an F test which factored out the effects of the two different E-pre substances.

The hypothesis which generated the experimentation was confirmed: Preference for the E-pre flavor was reliably higher when E-pre-X was novel than when it was familiar ($F = 6.42$, $1/44$ df, $p < .01$, one tail). The novel E-pre-X treatment did not produce complete interference because it yielded lower preferences for E-pre than the no E-post (no toxicosis) treatment ($F = 10.77$, $1/43$ df, $p < .002$); that is, some association of E-pre with E-post occurred even when E-pre-X was novel. The reduction in interference produced by latent inhibition was so pronounced that there was no reliable difference between the familiar E-pre-X treatment and the no E-pre-X treatment ($F = 0.27$, $1/43$ df, $p > .30$, one tail); that is, an almost maximal aversion to E-pre was obtained although the familiar E-pre-X was consumed between E-pre and toxicosis.

E. IMPLICATIONS

1. Confirmation of Concurrent Interference Theory

The preceding results strongly confirm concurrent interference theory. The amount of interference produced by E-pre-X was shown to be an increasing function of the strength of its association with E-post. Since such concurrent interference does occur, it follows that if a large number of E-pre-X's capable of becoming strongly associated with E-post were to occur during a delay, long-delay learning would be impossible. This confirms the earlier conjecture that long-delay learning actually occurs partly because of the lack of concurrent interference. There is no support in these results for any theory which asserts that long-delay learning occurs because the usual rules of associative learning are not applicable to flavor–toxicosis associations. Admittedly, the preceding results are all based on the case in which E-pre-X is a flavored substance and it can be validly claimed that this logically places a potential limit on their generality. Nevertheless, the results are still strong enough that the burden of proof is now on those who would dispute the generality of concurrent interference theory.

2. Speculations about the Rat as a Logician

If E-pre's and E-post's are considered as potential causes and effects, the concurrent interference principle makes the rat act as if it adheres to a sophisticated probabilistic notion of causality. If it can attribute an effect to an earlier cause with near certainty on the basis of criteria of relevance, psychophysical intensity, and its past experience, the rat considers it unlikely that some other earlier cause could also have produced the effect. If the effect cannot be attributed to any one cause with near certainty, the rat keeps a more open mind. Since the concurrent interference principle makes the rat behave so cleverly, it would seem ill-advised to attribute the occurrence of concurrent interference to a deficit in the animal's ability to form associations or to retain information. It seems better to consider it a biological adaptation to an environment in which a probabilistic notion of causality has good predictive power and in which indiscriminate associations would be maladaptive.

The rat's failure to associate over delays of more than a few seconds in the traditional experimental situations of animal learning should not diminish our newfound respect for its intelligence. In these situations, a large number of events occur prior to a later event and the concern is whether a particular, arbitrarily selected event among these will become associated with the later event. Without any reasonable criterion for selective association, the problem would be insoluble for any animal, no matter how intelligent.

VII. Summary

It is now certain that rats can associate directly between two events (called E-pre and E-post) separated by a long temporal delay. This disproves the traditional notion that temporal contiguity (or near contiguity) is a necessary condition for learning. The purpose of this chapter is to supply a substitute for this traditional notion. A single set of theoretical concepts was developed to explain both long-delay learning and the many findings which had led to the traditional belief that long-delay learning is impossible without mediation.

The most important concept in the theory is concurrent interference. This occurs when the events of the reference association become associated with other events which are not part of the reference association. For instance, if an event other than E-pre becomes associated with E-post, it will produce concurrent interference. Concurrent interference is assumed to prevent the reference association from being learned. Concurrent interference differs from proactive and retroactive interference in that the interference occurs during training instead of prior to training or after it. Substantial evidence for the existence of concurrent interference was cited.

In the traditional learning experiments in which association over even a short delay is difficult to obtain, the delay contains many events which produce concurrent interference by becoming associated either with E-pre or with E-post. For instance, a rat subjected to delayed reinforcement in a runway is bound to emit many movements during the delay which may become associated with the reinforcement. By the process of concurrent interference, these extraneous associations tend to prevent the running response from becoming associated with the reinforcement. It was shown that this approach can account for delay of reinforcement data which Spence explained in terms of a more traditional approach.

In the experimental situations in which learned associations over long delays readily occur, concurrent interference does not occur in strength due to a mechanism called relevance. When relevance is operative, E-pre and E-post are of such a nature that they readily become associated with each other and, what is more important, cannot readily become associated with the events which might usually occur during a delay. For instance, in the case of a flavor–toxicosis association, the flavor does not readily become associated with the many auditory, visual, and proprioceptive events which are likely to occur during the delay; nor can these intervening events readily become associated with toxicosis. Experiments by Garcia and his co-workers which gave rise to this conjecture were described. It also was shown that when flavored substances other than E-pre are consumed between the E-pre substance and toxicosis, they tend to prevent the reference association from developing. Since the introduction of concurrent interference into situations where long-delay learning normally occurs tends to prevent it from occurring, these results support the conjecture that lack of substantial concurrent interference is largely responsible for long-delay learning.

REFERENCES

Baker, T. W. Properties of compound conditioned stimuli and their components. *Psychological Bulletin,* 1968, **70,** 611–625.

Bersh, P. J. The influence of two variables upon the establishment of a secondary reinforcer for operant responses. *Journal of Experimental Psychology,* 1951, **41,** 62–73.

Bixenstine, V. E. Secondary drive as a neutralizer of time in integrative problem solving. *Journal of Comparative and Physiological Psychology,* 1956, **49,** 161–166.

Braveman, N., & Capretta, P. J. The relative effectiveness of two experimental techniques for the modification of food preferences in rats. *Proceedings of the 73rd Annual Convention of the American Psychological Association.* Washington, D.C.: American Psychological Association, 1965. Pp. 129–130.

Breland, K., & Breland, M. The misbehavior of organisms. *American Psychologist,* 1961, **16,** 681.

Campbell, C. S. The development of specific preferences in thiamine-deficient rats: Evidence against mediation by aftertastes. Unpublished master's thesis, University of Illinois at Chicago Circle, 1969.

Capaldi, E. J. A sequential hypothesis of instrumental learning. In K. W. Spence & J. T. Spence (Eds.), *The psychology of learning and motivation.* Vol. 1. New York: Academic Press, 1967. Pp. 67–157.

Capaldi, E. J., & Spivey, J. E. Intertrial reinforcement and aftereffects at 24-hour intervals. *Psychonomic Science,* 1964, **1,** 181–182.

Capretta, P. J. An experimental modification of food preferences in chickens. *Journal of Comparative and Physiological Psychology,* 1961, **54,** 238–242.

Capretta, P. J., & Moore, M. J. Appropriateness of reinforcement to cue in the conditioning of food aversions in chickens (*Gallus gallus*). *Journal of Comparative and Physiological Psychology,* 1970, **72,** 85–89.

Carlton, P. L., & Vogel, J. R. Habituation and conditioning. *Journal of Comparative and Physiological Psychology,* 1967, **63,** 348–351.

Deese, J., & Hulse, S. H. *The psychology of learning.* (3rd ed.) New York: McGraw-Hill, 1967.

Dietz, M. N., & Capretta, P. J. Modification of sugar and sugar-saccharin preference in rats as a function of electrical shock to the mouth. *Proceedings of the 75th Annual Convention of the American Psychological Association.* Washington, D.C.: American Psychological Association, 1967. Pp. 161–162.

Farley, J. A., McLaurin, W. A., Scarborough, B. B., & Rawlings, T. D. Pre-irradiation saccharin habituation: A factor in avoidance behavior. *Psychological Reports,* 1964, **14,** 491–496.

Ferster, C. B., & Hammer, C. Variables determining the effects of delay in reinforcement. *Journal of the Experimental Analysis of Behavior,* 1965, **8,** 243–254.

Garcia, J. The faddy rat and us. *New Scientist and Science Journal,* 1971, **49,** 254–256.

Garcia, J., & Ervin, R. R. A neuropsychological approach to appropriateness of signals and specificity of reinforcers. *Communications in Behavioral Biology,* 1968, Part A, **1,** 389–415.

Garcia, J., Ervin, F. R., Yorke, C. H., & Koelling, R. A. Conditioning with delayed vitamin injections. *Science,* 1967, **155,** 716–718.

Garcia, J., Kimeldorf, D. J., & Koelling, R. A. Conditioned aversion to saccharin resulting from exposure to gamma radiation. *Science,* 1955, **122,** 157–158.

Garcia, J., & Koelling, R. A. Relation of cue to consequence in avoidance learning. *Psychonomic Science,* 1966, **4,** 123–124.

Garcia, J., & Koelling, R. A. A comparison of aversions induced by X-rays, toxins, and drugs in the rat. *Radiation Research,* 1967, **7,** 439–450.

Garcia, J., McGowan, B. K., Ervin, F. R., & Koelling, R. A. Cues: Their effectiveness as a function of the reinforcer. *Science,* 1968, **160,** 794–795.

Gorry, T. H., & Ober, S. E. Stimulus characteristics of learning over long delays in monkeys. Paper delivered at the 10th annual meeting of the Psychonomic Society, San Antonio, November, 1970.

Guthrie, E. R. *The psychology of learning.* (Revised) New York: Harper, 1952.

Harker, G. S. Delay of reward and performance of an instrumental response. *Journal of Experimental Psychology,* 1956, **51,** 303–310.

Hilgard, E. R., & Bower, G. H. *Theories of learning.* (3rd ed.) New York: Appleton, 1966.

Holman, G. L. The intragastric reinforcement effect. *Journal of Comparative and Physiological Psychology,* 1969, **69,** 432–441.

Honig, W. K. Attention and the modulation of stimulus control. In D. Mostofsky (Ed.), *Attention: Contemporary studies and analyses.* New York: Appleton, 1970. Pp. 193–238.

Kamin, L. J. Selective association and conditioning. In N. J. Mackintosh and W. K. Honig (Eds.), *Fundamental issues in associative learning*. Halifax: Dalhousie Univ. Press, 1969. Pp. 42–64.

Kelleher, R. T. Chaining and conditioned reinforcement. In W. K. Honig (Ed.), *Operant behavior: Areas of research and application*. New York: Appleton, 1966. Pp. 160–212.

Kelleher, R. T., & Gollub, L. R. A review of positive conditioned reinforcement. *Journal of the Experimental Analysis of Behavior*, 1962, 5, 543–597.

Kimble, G. A. *Hilgard and Marquis' Conditioning and Learning*. (2nd ed.) New York: Appleton, 1961.

Konorski, J. *Integrative activity of the brain: An interdisciplinary approach*. Chicago: Univ. of Chicago Press, 1967.

Lubow, R. E. Latent inhibition: Effects of frequency of nonreinforced preexposure of the CS. *Journal of Comparative and Physiological Psychology*, 1965, 60, 454–457.

Mackintosh, N. J., & Honig, W. K. Blocking and enhancement of stimulus control in pigeons. *Journal of Comparative and Physiological Psychology*, 1970, 73, 78–85.

McLaurin, W. A., Farley, J. A., & Scarborough, B. B. Inhibitory effect of preirradiation saccharin habituation on conditioned avoidance behavior. *Radiation Research*, 1963, 18, 473–478.

Nachman, M. Learned taste and temperature aversions due to lithium chloride sickness after temporal delays. *Journal of Comparative and Physiological Psychology*, 1970, 73, 22–30.

Pavlov, I. P. *Conditioned reflexes*. (Trans. by G. V. Anrep.) London and New York: Oxford Univ. Press, 1927.

Petrinovich, L., & Bolles, R. C. Delayed alternation: Evidence for symbolic processes in the rat. *Journal of Comparative and Physiological Psychology*, 1957, 50, 363–365.

Petrinovich, L., Bradford, D., & McGaugh, J. L. Drug facilitation of memory in rats. *Psychonomic Science*, 1965, 2, 191–192.

Rescorla, R. A. Conditioned inhibition of fear. In N. J. Mackintosh and W. K. Honig (Eds.), *Fundamental issues in associative learning*. Halifax: Dalhousie Univ. Press, 1969. Pp. 65–89.

Revusky, S. H., & Bedarf, E. W. Association of illness with prior ingestion of novel foods. *Science*, 1967, 155, 219–220.

Revusky, S. H., & Garcia, J. Learned associations over long delays. In G. H. Bower (Ed.), *The Psychology of learning and motivation*. Vol. 4. New York: Academic Press, 1970. Pp. 1–84.

Reynolds, G. S. Attention in the pigeon. *Journal of the Experimental Analysis of Behavior*, 1961, 4, 203–208.

Rozin, P. Central or peripheral mediation of learning with long CS-UCS intervals in the feeding system. *Journal of Comparative and Physiological Psychology*, 1969, 67, 421–429.

Seligman, M. E. P. On the generality of the laws of learning. *Psychological Review*, 1970, 77, 406–418.

Shilling, M. An experimental investigation of the effect of a decrease in the delay of reinforcement upon instrumental response performance. Unpublished master's dissertation, State University of Iowa, 1951.

Solomon, R. L., & Wynne, L. C. Traumatic avoidance learning: Acquisition in normal dogs. *Psychological Monographs*, 1953, 67 (No. 4, Whole No. 354).

Spence, K. W. The role of secondary reinforcement in delayed reward learning. *Psychological Review*, 1947, 54, 1–8.

Spence, K. W. *Behavior theory and conditioning*. New Haven, Conn.: Yale Univ. Press, 1956.

Sutherland, N. S., & Mackintosh, N. J. Discrimination learning: Non-additivity of cues. *Nature (London)*, 1964, **20**, 528–530.

Ulrich, R., & Azrin, N. H. Reflexive fighting in response to aversive stimulation. *Journal of the Experimental Analysis of Behavior*, 1962, **5**, 511–520.

Wilcoxon, H. C., Dragoin, W. B., & Kral, P. A. Differential conditioning to visual and gustatory cues in quail and rat: Illness induced aversions. *Psychonomic Science*, 1969, **17**, 52. (Abstract)

Wittlin, W. A., & Brookshire, K. H. Apomorphine-induced conditioned aversion to a novel food. *Psychonomic Science*, 1968, **12**, 217–218.

Zahorik, D. M., & Maier, S. F. Appetitive conditioning with recovery from thiamine deficiency as the unconditioned stimulus. *Psychonomic Science*, 1969, **17**, 309–310.

Chapter 5

MODIFICATION OF MEMORY
STORAGE PROCESSES[1]

James L. McGaugh and Ronald G. Dawson

Of the variety of mechanisms of adaptation, one of the most important is the capability of the individual organism to learn. Such adaptation as a result of environmental encounters requires that an animal be able to store experiences, retrieve the stored information, and perform responses which are appropriate in terms of the experience. Unfortunately, in spite of thousands of experimental studies and dozens of theories, we do not as yet know very much about how animals store, retain, and retrieve information; and we know virtually nothing about processes underlying the control of performance of learned responses. Obviously, complete understanding of the nature of memory in animals will require detailed investigation of the characteristics as well as the mechanisms underlying each of these abilities.

I. Neurobiological Correlates of Learning and Memory:
Problem of Permanence

Our theoretical conceptions of learning and memory have been particularly weak on the question of the permanence of memory. This state of affairs seems to stem directly from the vagueness of our definitions of learning. Does the requirement of "a relatively permanent change," which is included in most definitions of learning, imply that all memories are permanent or that some are transient while others are not? On this question, the theories of learning are usually silent. There is, of course, no *a priori* reason to assume that all consequences of experience are permanent. In fact, there appear to be many reasons why this should not be the case. Many of the stimuli surrounding an

[1] The preparation of this review was supported by research grant MH 12526, Predoctoral Traineeship MH 11095 and Predoctoral Fellowship MH 44722 from the National Institute of Mental Health, United States Public Health Service. We thank Karen Dodd for her assistance in preparation of the manuscript.

animal have little or no adaptive significance. It makes little sense in terms of economy or adaptation that all stimuli should produce enduring consequences in the brain. On the other hand, it is of adaptive value to be able to store some long-term representation of experiences that are either particularly meaningful (in terms of their consequences) or frequently repeated. We will need to understand the conditions under which memories are transient, as well as those leading to the formation of lasting or long-term memory, if we are to have a complete understanding of the neurobiological bases of memory.

A wealth of neurophysiological evidence (John, 1967) and biochemical evidence (e.g., von Hungen, Mahler, & Moore, 1968) suggests that the brain typically exhibits transient changes to environmental stimulation. Patterns of brain electrical activity are continually in flux, and in a learning situation they may change drastically and then disappear. Further, at the biochemical level Lajtha and Toth (1966) have estimated that approximately 90% of cerebral proteins show an average half life of between 10 and 20 days. Cytoplasmic RNA in the cortex has a half life of approximately 12 days, and brain ribosomes turn over as units with approximately the same regularity (i.e., 12 days) (von Hungen *et al.*, 1968). However, several studies suggest that other cerebral constituents including lipids, nucleic acids, and carbohydrates as well as some proteins turn over much more slowly (Thompson & Ballou, 1956). Thus, the search for neurobiological correlates of memory must be guided by the evidence that there are candidates which are "relatively transient," as well as those which are "relatively permanent." If a neurobiological change is to serve as the basis of memory it must, of course, vary with the *behaviorally* measured memory (cf. Dingman & Sporn, 1964). That is, the neurobiological changes underlying memory must (1) be present after learning, (2) remain as long as retention is evidenced, and (3) disappear when the effects of learning are no longer demonstrable. Obviously, these are difficult criteria to satisfy. Again, the difficulty is that the search for such correlations must not be restricted by the assumption that all learning necessarily results in long-term neural changes.

Recently, several investigators have reported finding neurobiological correlates—both chemical and electrophysiological—of training (cf. Glassman, 1969; John, 1967). However, none of the correlates has met the three criteria listed above and, consequently, it cannot yet be concluded that the correlates are involved in processes underlying memory. There are several reasons for this. First, the neurobiological changes (e.g., altered RNA synthesis, modified patterns of electrophysiological activity) might be due to processes involved in the reception of information and the performance of responses rather than to the formation of memory. Second, it could be that in some cases the changes are due to alterations in motivational state (e.g., arousal) produced by the training experience. Third, it may be impossible to obtain the evidence essential for deciding whether the criteria have been met. For example, if an experience produces both transient (i.e. short-term memory) effects and durable (i.e., long-term memory) effects, the transient neurobiological change underlying short-term memory would be rejected as a basis of memory according to the criteria above simply because the behavioral evidence of memory (i.e., long-term memory) outlasts the neurobiological transient. In view of these difficulties, it seems unlikely that an understanding of the neurobiological bases of memory will result *solely* from correlational studies.

II. Experimental Analysis of Time-Dependent Processes of Memory Storage

An alternative method of approaching the problem of the bases of memory, and one currently employed in numerous laboratories, is that of attempting to modify memory processes by administering treatments which affect central nervous system functioning. This approach is based on the assumption that an understanding of the nature of the effect of the treatment on memory (i.e., amnesia or facilitation of learning), considered together with knowledge of the mechanisms of action of the treatment, might provide important clues to the neurobiological process underlying memory. In the remainder of this chapter we shall examine the results and implications of some of these investigations. We shall conclude by examining several models of memory storage growing directly out of this research, which we feel may prove to be important tools for future research.

Most of the research in this area has been guided by some form of a "dual-trace" hypothesis, that is, the assumption that long-term memory (LTM) and short-term memory (STM) are based on different processes (e.g., Hebb, 1949; Gerard, 1949). Although there is some purely behavioral evidence bearing on the "dual-trace" hypothesis, most of the experiments concerning time-dependent processes in memory storage have used experimental procedures which modify neural activity in some way. Generally the subjects in such experiments are first trained on a task and are then treated either immediately or at some later time. Then, at some time after the animals have recovered from the acute effects of the treatment, retention tests are given. Inferences concerning the effects of the treatment on memory storage processes are based on retention performance. Under ideal circumstances the animals are neither trained nor tested while under the acute effects of the treatments. Figure 1 shows the critical time periods in experiments of this kind. Treatments administered at points A and E may be used to assess the proactive effects of the treatment upon training and retest performance respectively. Treatments administered at points B, C, and D may be used to assess the temporal gradient of effectiveness of the treatment, without the confounding of performance effects, since subjects are neither trained nor tested under the influence of the treatment. It is of historical interest to note

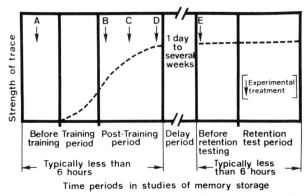

Fig. 1. Times of experimental treatment in studies of memory storage. (From McGaugh, 1969.)

that the guiding theory (Hebb, 1949) and the experimental procedure involving posttrial treatments (Duncan, 1949) were both published in the same year two decades ago.

By far, most of the research in this area has involved treatments designed to *interfere* with memory storage. Electroconvulsive shock (ECS) is the technique used most frequently. But numerous other techniques have been used: spreading depression, depressant drugs, convulsant drugs, antibiotics, electrical stimulation of the brain, hypoxia, hypothermia, and brain lesions (Agranoff, Chapter 6; Barondes, 1968; Glickman, 1961; McGaugh, 1966; McGaugh & Herz, 1971). In general, all of these techniques have been reported to produce retrograde amnesia (RA). That is, the treatments administered after training produce impairment in memory, and the degree of impairment decreases as the time between training and treatment is increased, such that at some treatment delay there is no further impairment of memory.

It is also possible to facilitate learning by treating animals with drugs or electrical stimulation of the brain. The degree of facilitation obtained depends upon many factors including the strain of animals, drug, drug dose, and type of training task. As might be expected according to the evidence from "disruption" studies, the degree of facilitation of learning obtained with such treatments decreases as the interval between training and drug injection or electrical stimulation is increased and at some point there is no further facilitation of learning (McGaugh, 1968b, 1969; McGaugh & Herz, 1971; McGaugh & Petrinovich, 1965).

Since both facilitative and disruptive treatments lose their effectiveness as the interval between training and treatment increases, it has been argued that the treatments act by affecting processes initiated by the training, which underly memory consolidation, rather than by influencing performance on the subsequent retention trials.

A conventional interpretation of these findings is that the treatments used in these studies act in some way to impair or facilitate the neurobiological processes involved in the formation of LTM. Further, it is assumed by many investigators that the gradient of RA provides a measure of "consolidation time"; i.e., the time required for formation of LTM is indexed by the time at which facilitative and disruptive treatments are no longer effective in altering retention. Such evidence would have obvious value in aiding in the specification of the time course of biological substrates which might underly LTM formation.

These conventional conclusions and interpretations have recently been seriously challenged on both theoretical and empirical grounds. Many recent findings are difficult to interpret in conventional terms. Consequently, several alternative hypotheses of the bases of the effects of treatments which affect learning when administered after training have been proposed. Most of the very recent research in this area has centered on the following issues: (a) the permanence of the treatment effects, (b) the nature of the treatment gradient, (c) time vs. events as causes of the treatment gradient, (d) the number of and interrelationships among inferred memory processes, and (e) the neurobiological bases of the treatment effects. Although major controversies have concerned the nature of the effects found with ECS, the arguments which prove to be of substance apply over a wide range of other treatment conditions.

Each of these problems is discussed briefly (but unfortunately not solved) in the following sections. The problems are as complex as they are controversial, and the discussions can serve only to focus the questions.

III. Permanence of Treatment Effects

According to the conventional interpretation of studies of disruption of memory storage, the amnesia should be permanent. If ECS and other treatments produce amnesia by blocking storage processes, memory should not return when the animals recover from the acute effects of treatments. Under some conditions we have found this to be the case (e.g., Luttges & McGaugh, 1967; Zornetzer & McGaugh, 1969). For example, in the study by Luttges and McGaugh (1967) mice received a foot shock (FS) as they stepped from a small platform through a small hole and onto a metal floor. Some mice then received an ECS within a few seconds. Other mice received only a FS or an ECS, or no treatment. All mice were subsequently retested at one of three retention intervals (12 hours, 1 week or 1 month later). On the retest, most of the mice in groups which received FS and no ECS would not leave the small platform. Their response latencies were higher than those of any of the other groups (Fig. 2). Similar retention performance was found at all retention intervals. Memory for the FS experience appears to be stable over the 1-month period. The ECS treatment produced significant and long-lasting amnesia. Animals given FS and ECS stepped down

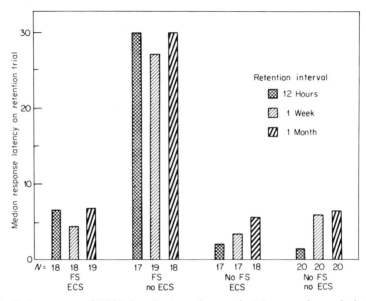

Fig. 2. Permanence of ECS-induced retrograde amnesia. Mice were given a single trial in a step-through inhibitory avoidance task and treated with FS and ECS as indicated. Independent groups were given retention tests 12 hours, 1 week, or 1 month later. (From Luttges & McGaugh, 1967.)

almost as quickly as did unpunished controls on all retention tests. The latencies in groups tested 1 week or 1 month after training were as low as those tested within 12 hours. Thus, amnesia was stable over at least a 1-month period. Similar results have been obtained by other investigators (Chevalier, 1965; Geller & Jarvik, 1968a; Greenough, Schwitzgebel, & Fulcher, 1968; Herz & Peeke 1968).

However, the findings of some studies suggest that the memory impairment found with ECS is only temporary (e.g., Nielson, 1968; Zinkin & Miller, 1967). Weiskrantz (1966) and Nielson (1968) have proposed somewhat similar interpretations of the recovery effect. Both hypotheses assume that ECS produces changes in brain activity or brain "state" which last for several days. Essentially, their hypotheses argue against the use of ECS-produced disruption gradients as support for the notion that LTM formation requires a considerable amount of time. The hypotheses assume that in the case of ECS (and presumably, but not explicitly, other amnesic agents as well) amnesia caused by treatments given after training is caused *not* by interference with storage (except for treatments given within a few seconds) but rather by interference with retrieval. Retrieval is assumed to take place when the neural signal underlying memory is sufficiently strong to exceed the background level or noise of ongoing brain activity. Weiskrantz has suggested that the treatment decreases the signal-to-noise ratio by elevating background noise levels in the brain

(Weiskrantz, 1966; see Fig. 10). Nielson assumes that ECS elevates the threshold for the elicitation of the learned responses by altering brain state (Nielson, 1968). Both hypotheses predict that the memory impairment will decrease as the treatment effects subside with time, that is, as the brain returns to its "normal" state. These hypotheses do not, of course, explain the permanence of amnesia seen in most experiments. Beyond that, these hypotheses do not explain the retrograde amnesic effect of amnesic agents. Typically, greatest amnesia is obtained if the ECS is given shortly after training. If noise produced by the treatment decays over time then the degree of amnesia produced by ECS might be expected to increase as the interval between training and treatment is increased, if the interval between training and retention testing remains constant. That is, there should be greater impairment of retrieval if the ECS treatments are given closer in time to the retention tests. Moreover, these "neural noise" or "brain state" hypotheses are silent with respect to the findings of retrograde facilitation of memory storage. Thus, at best, such hypotheses are incomplete.

Beyond that, however, explicit tests of some of the implications of these hypotheses have failed to confirm predictions derived from them. For example, if animals are first trained on a task and are then given an ECS treatment shortly (i.e., 30 minutes) prior to the retention test, the ECS treatment has little if any effect on the retention performance (McGaugh & Longacre, 1969; Zerbolio, 1969). In the study by McGaugh and Longacre (1969) mice were highly trained in the performance of an active–avoidance response in a shuttle-box situation. Subsequently ECS was administered 2 hours, 1 hour, 15 minutes, or immediately prior to a retention test. Half of the animals in these groups were allowed to convulse and in half, the convulsions were prevented by lightly anesthetizing the animals prior to ECS. ECS-induced convulsions occurring 1 hour or less prior to the retention tests produced a slight impairment in performance. However, when convulsions were prevented with ether there was no impairment even if the ECS was administered immediately prior to retesting. Thus, proactive impairment of performance when it occurs is not due simply to the passage of current through the brain. In direct contrast, the retrograde amnesic effects of ECS appear to be so induced since ECS produces retrograde amnesia even if the behavioral convulsion is prevented by light ether (McGaugh & Alpern, 1966; McGaugh & Zornetzer, 1970; Zornetzer & McGaugh, 1971; McGaugh, Dawson, Coleman & Rawie, 1971). However, the proactive effects of ECS on acquisition might be different from proactive effects of ECS on retention of a well-learned response.

In a direct test of Nielson's "state-dependency" interpretation of ECS impairment of memory (1968), Zornetzer and McGaugh (1969) failed to find evidence for the two major predictions of this hypothesis. Mice in one group were trained on a one-trial inhibitory avoidance task under a normal brain state and given ECS immediately after training. These mice showed amnesia whether the retention test was given 24 hours or 96 hours following training. There was no evidence of recovery of memory under these conditions, even though animals would, according to Nielson, have gradually returned to a normal brain state over the 96-hour retention interval.

Mice in another group were placed on a grid floor and given a "noncontingent" FS (that is, the FS was simply administered—no response was required) followed immediately by an ECS. According to Nielson, this treatment should produce an alteration in brain state lasting for several days. These mice were then trained on the one-trial inhibitory avoidance task 24 hours later and given an ECS immediately after the training trial. A retention test was given 24 hours later. Thus, both the training trial and the retention test trial were given 24 hours after an ECS treatment. According to Nielson's hypothesis, the animals should show good retention of the avoidance response on the retention test because the "brain state" at the time of the retention test is similar to the "brain state" during the original training. The results did not support Nielson's hypothesis. The animals showed no evidence of memory on the retention test.

Other recent experiments have examined the possibility that the memory of an event which is followed by ECS may be stored in the "brain state" which is produced by ECS (W. J. Hudspeth, personal communication; Reichert, 1968). Furthermore, the decrease in treatment effectiveness with progressively longer intervals between training and treatment may reflect the relative proportions of storage which occur in the normal as opposed to the ECS-induced brain state, rather than ECS-produced disruption of a progressively more stable memory trace. According to this hypothesis it might be possible to restore memory in amnesic animals by giving an ECS to the animals shortly before the retention test, that is, by reinstating the brain state that existed during the posttraining period. To examine this possibility (McGaugh & Landfield, 1971) mice were given a single training trial in the inhibitory avoidance task followed immediately by an ECS. On the following day, half of the animals were given an ECS 1 hour before the retention test and half were untreated. Neither group showed evidence of retention. Thus, our experiments have provided no support for any of the various forms of the brain "state-dependency" hypothesis. More generally, we have found no conditions under which the amnesia produced by ECS is only temporary.

Other investigators, however, have isolated conditions under which RA may be only temporary. For example, Pagano, Bush, Martin, and Hunt (1969) found recovery from amnesia with very low intensities of ECS (i.e., just sufficient to convulse the animal) but no recovery with high intensities. The conditions under which RA is only temporary cannot, of course, be used to make inferences concerning storage processes but could be very important for analysis of retrieval processes. However, there are many conditions under which stable retrograde amnesia can be obtained, and it is only these conditions which can provide information critical for memory storage hypotheses.

IV. Nature of the Treatment Gradient

Estimates of "consolidation" time based on studies using posttraining treatments have ranged from 10 seconds to 24 hours. Most estimates are within the range of minutes to hours. Since RA and facilitation gradients are based on many different treatments and a variety of training conditions, it is not surprising that the estimates vary. However, even with the same species of mice, trained on relatively similar inhibitory avoidance tasks, RA gradients produced by ECS have been shown to vary from minutes to hours. The RA gradient observed has been shown to depend upon several factors including the nature of the current (Alpern & McGaugh, 1968; Dorfman & Jarvik, 1968a; Miller, 1968), the route of application (Dorfman & Jarvik, 1968b; Ray & Barrett, 1969) as well as the animals' state during the training–treatment interval (Fishbein, McGaugh, & Swarz, 1971) and the time of day at which the treatment is given (Stephens & McGaugh, 1968a, 1968b; Stephens, McGaugh, & Alpern, 1967). Clearly, under these restricted conditions RA curves have been shown to be curves of susceptibility, not curves of "consolidation". The nature of the RA gradient obtained depends both upon the processes which are susceptible to modification as well as the characteristics of the treatment used to assess them.

Cherkin (1969) has shown very clearly that in experiments with chicks using the drug flurothyl as the amnesic treatment the RA gradients obtained depend upon the concentration of the drug (which is inhaled) and the duration of the treatment. With low concentrations and short durations the RA gradient is quite short. Perhaps the most important point is that even with high concentrations and long durations, the treatment effectiveness is still graded; only marginal effects are obtained with 24-hour posttraining treatments. Such findings imply that while all treatments reflect an underlying process, or processes which become less susceptible over time, the effect of the treatment upon such processes can be increased as the treatment strength is increased. If we are accurately to assess the extent to which these processes have progressed when the treatment is administered, we must examine the nature of the disruptive effects of a wide variety of treatment strengths. Do strong treatments stop consolidation or even induce decay while weak treatments merely slow it down? It is only by assessing retention at various intervals following the treatment that we can answer these questions. An assessment of the "true" consolidation gradient for any task would seem to require that treatments which arrest consolidation are selected, such that the level of retention measured at retest might be used to estimate the level of consolidation up to the time that the treatment was given. This assumption demands that the retention measured at various intervals following treatment remains constant.

Several investigators have suggested that the wide range of RA gradients obtained in inhibitory avoidance tasks are due, not to a unitary disruption of storage, but rather to two qualitatively different events. The short RA gradient of approximately 30 seconds (e.g., Chorover & Schiller, 1965, 1966) is assumed to reflect "true" amnesia. The long RA gradients are assumed to reflect not the decreasing susceptibility to disruption of memorial events, but rather an increase in the strength (incubation) of the avoidance response (Pinel & Cooper, 1966; Spevack & Suboski, 1969). Furthermore, incubation of an experimentally produced conditioned emotional response (CER) is presumed to underly the change in avoidance response strength. The degree of apparent retention is assumed to reflect the strength of the CER and ECS is assumed to impair "retention" by arresting the incubation of the CER (Spevack & Suboski, 1969). This last assumption clearly demands that the incubation gradient observed in animals tested at various intervals shortly following training parallels the RA gradient obtained when animals are given ECS at the same intervals and tested at a later time. This hypothesis is supported by some evidence. Some studies have found that, in inhibitory avoidance learning, the probability of avoidance on a retention test increases as the interval between training and testing is lengthened. Pinel and Cooper (1966) have shown that under some conditions the curves based on retention performance of animals tested at various times following training correspond rather closely to the retrograde amnesic gradient obtained when animals are given ECS at one of several times after training and tested 24 hours later. The change in retention performance with time is assumed to be due to incubation of the avoidance response.

As Dawson (1971) has pointed out, the assumption that the short RA gradient (i.e., less than 30 seconds) represents the "true" memorial effect of ECS appears to result from the selection of an arbitrary low (30 seconds) cut-off latency for responding during testing in the inhibitory avoidance situation (cf. Schneider, Kapp, Aron, & Jarvik, 1969). Further, this assumption is clearly incompatible with the extensive evidence showing systematic graded effects of treatments given up to several hours after training (Alpern & McGaugh, 1968; Dorfman & Jarvik, 1968a; Miller, 1968). Second, there is little evidence that a CER is produced in the inhibitory avoidance paradigms typically used to assess ECS-produced RA, or that a single ECS affects the development of a CER when produced. Third, long RA gradients have been produced in appetitive paradigms (Tenen, 1965a, 1965b) where animals are first trained to approach food rather than avoid a punishing experience. Consequently, the question of CER production does not arise. Finally, the assumption that ECS halts CER incubation is clearly incompatible with the majority of studies which have compared incubation curves with RA curves. We have found (cf. McGaugh, 1966) that the retention test performance in animals given only a training trial at most intervals (up to 1 hour) following training is better than that of animals given only an ECS at that interval and a retention test 24 hours later. Thus the "incubation" curve is not necessarily identical with or even parallel to the RA curve. In one-trial aversive learning with chicks, Cherkin (1969) has found long

RA gradients with flurothyl even though performance is high immediately after training in untreated animals. Pinel (1969) obtained similar findings with rats given a single training trial on an appetitive task. Thus, although performance on some tasks may change with time following training, the nature of the change appears to be unrelated to the characteristics of the RA gradient.

On the basis of available evidence, RA curves seem best interpreted as indicating that memory storage processes become decreasingly susceptible to interference with time following training. Generally, little or no interference is found when the interval between training and treatment exceeds several hours. Memory storage is not complete at the time of an experience. The processes initiated by training remain labile for a fairly long interval following training.

V. Drug Facilitation of Memory Storage

The conclusion that memory storage processes are active for fairly long intervals following an experience is also supported by studies of facilitation of learning. In a series of experiments conducted over the past 12 years, it has been shown that in a wide variety of tasks the learning of animals is enhanced by

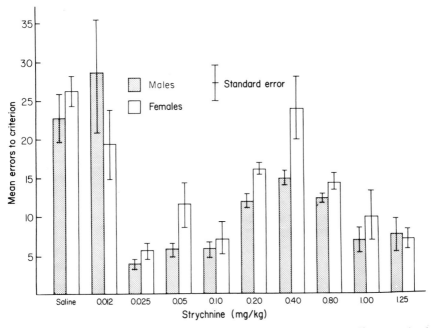

Fig. 3. The effect of posttraining administration of strychnine sulfate on visual discrimination learning in mice. Mean number of errors made by saline control groups and experimental groups (N/subgroup = 6) receiving a range of drug doses are shown. The dose response curve is clearly biphasic. Greatest facilitation can be seen with lower doses (0.25–0.10 mg/kg, ip) and higher doses (1.00–1.25 mg/kg, ip). Smaller effects can be seen with intermediate doses (0.20–0.80 mg/kg, ip). (From McGaugh, 1968b.)

administration of CNS stimulants shortly after training [see McGaugh (1968b) and McGaugh and Petrinovich (1965) for reviews of this area]. Facilitating effects have been found with several drugs including strychnine, picrotoxin, pentylenetetrazol, caffeine, amphetamine, nicotine, diazadamantanol, and physostigmine (e.g., Franchina & Moore, 1968; Garg & Holland, 1968; Oliverio, 1968; Stratton & Petrinovich, 1963; Westbrook & McGaugh, 1964).

In a series of experiments McGaugh and Krivanek (Krivanek & McGaugh, 1968, 1969; McGaugh & Krivanek, 1970) studied the effects of drug dose and time of drug administration on visual discrimination learning in mice In these studies, mice were trained to perform a black–white simultaneous visual discrimination in a Y-maze for food reward. Animals were given three trials each day until a criterion of 9 out of 10 correct responses was reached Intraperitoneal drug injections were administered at various times after each daily session. In the dose–response determinations, as with all placebo conditions, injections were administered immediately following the third daily trial. The times of other injections are indicated below. The dose–response results of studies using strychnine, pentylenetetrazol, and amphetamine are shown in Figs. 3, 4, and 5. Clearly, facilitation can be obtained with a wide range of doses of each drug.

Figures 6, 7, and 8 show the effects of time of injection, either pre- or posttrial, on learning. With each drug, the degree of facilitation decreased as the interval between training and posttrial injections was increased. Penty-lenetetrazol facilitated learning with posttrial injection intervals of 15 minutes but was ineffective with intervals of 30 minutes or longer. Strychnine facilitated learning with injections given up to 1 hour posttrial. Amphetamine facilitated learning when injected immediately posttrial but not when administered 15 minutes after training. These results indicate that memory storage processes are susceptible to facilitating influences for a relatively long period of time after training is terminated. This period, as indicated by the findings with strychnine, was at least 1 hour in length. With all drugs tested, the posttrial injection effect is time dependent. The gradient varies with the drug used to measure it.

Thus these findings complement those obtained with treatments that impair memory. Overall they provide strong evidence that memory storage processes remain labile for a relatively long period of time following a training experience. These findings suggest that the drugs facilitate learning by influencing the neurobiological processes underlying the consolidation or storage of long-term memory. Greatest facilitation is found if the drugs are administered shortly before or shortly after each drug training session. Little, if any, facilitation is found if the injections are given either 1 hour before daily training or 2 hours or longer after the training. Thus, the drugs appear to facilitate learning by enhancing the memory of experiences which occur either shortly after the injections or within an hour or so before the injections.

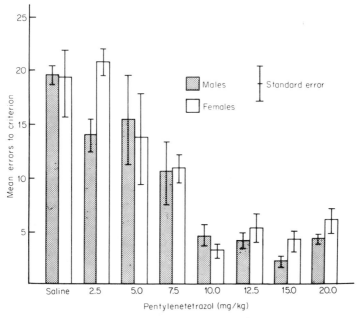

Fig. 4. The effect of posttraining administration of pentylenetetrazol on visual discrimination learning in mice. Mean number of errors made by saline control groups and experimental groups (N/subgroup = 6) receiving a range of drug doses are shown. The dose response effect is clearly graded. There is no significant facilitation with lower doses (2.5–5.0 mg/kg, ip). Degree of facilitation increased with increases in drug dose. (From Krivanek & McGaugh, 1968.)

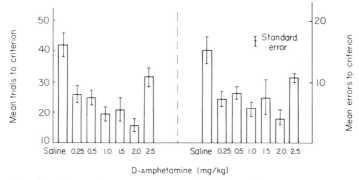

Fig. 5. The effect of posttraining administration of D-amphetamine on visual discrimination learning in mice. Mean number of errors made by saline control groups and experimental groups (N/group = 6) receiving a range of drug doses are shown. The dose–response curve appears to be U-shaped. Moderate facilitation was found with lower doses (0.25–0.5 mg/kg, ip); greatest facilitation was found with intermediate doses (1.0–2.0 mg/kg, ip); and only slight facilitation was found with the highest dose (2.5 mg/kg, ip) (From Krivanek & McGaugh, 1969).

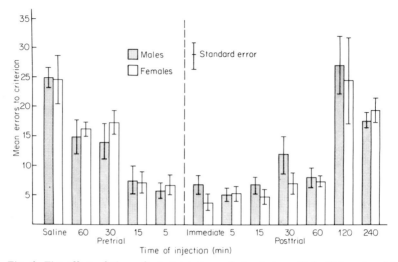

Fig. 6. The effect of time of administration of strychnine sulfate (0.1 mg/kg, ip) on mean errors to criterion on a visual discrimination learning task in mice. Pretrial injections of the drug produced the greatest facilitating effect when injected within 15 minutes of the trial. Facilitation is also obtained with posttrial injections up to 1 hour following training (N = 6/subgroup). (From McGaugh, 1968b.)

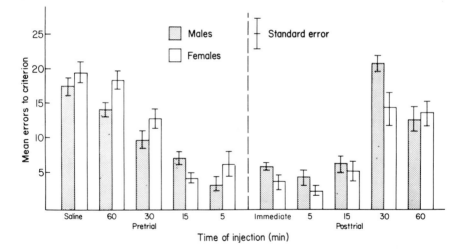

Fig. 7. The effect of time of administration of pentylenetetrazol (15 mg/kg, ip) on mean errors to criterion on a discrimination learning task in mice. Pretrial injections produced graded effects. The greatest facilitation was obtained with injections either 15 minutes or 5 minutes prior to training. Facilitation was also obtained with posttrial injections up to 15 minutes following training (N = 6/subgroup). (From Krivanek & McGaugh, 1968.)

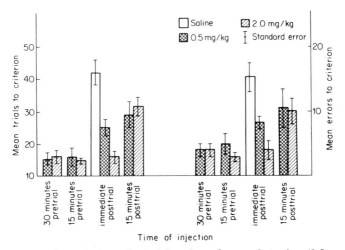

Fig. 8. The effect of time of administration of D-amphetamine (0.5 mg/kg and 2.0 mg/kg, ip) on mean errors and trials to criterion on a discrimination learning task in mice. Both measures of acquisition are essentially equivalent. Pretrial injections of the drug greatly facilitate learning (at both doses) when given at the 30 minute or 15 minute points. Posttrial injections show a graded facilitation; facilitation is greater immediately at 15 minutes. The graded effect is more pronounced for the high dose than for the low dose ($N = 6$/group). (From Krivanek & McGaugh, 1969.)

VI. Time vs. Events As Determinants of the Treatment Gradients

As we have indicated, the RA gradients have generally been interpreted as providing evidence that memory storage processes are time dependent. A number of recent studies have suggested, however, that this view may be oversimplified. For example, Davis and Agranoff (1966) and, more recently, Robustelli, Geller, and Jarvik (1968) have shown that susceptibility to amnesic agents is maintained if animals are kept in the training apparatus. Under such conditions the gradient of amnesia begins not with the time of training, but rather with the time of removal from the apparatus.

Schneider and Sherman (1968) have reported that RA in rats, following one-trial inhibitory avoidance training, can be obtained with long-training ECS intervals, providing that a second noncontingent FS precedes the ECS treatment. These findings have been interpreted as additional evidence that events during the training-treatment interval may render memory more susceptible to disruption and that the continuance of some consolidation process over time following learning is not sufficient to ensure memory stability. Moreover, they suggest an alternative explanation of the RA gradient, i.e., that the gradient reflects not the consolidation of memory but some nonassociative interaction between the consequences of reward or punishment (such as changes in level of arousal) in the learning situation and ECS. However, such an interpretation is clearly tentative. In a comparable study using mice, it was found that a FS given prior to ECS did not increase the effectiveness of ECS as an amnesic agent (McGaugh, Alpern, & Luttges, in preparation). Other studies using rats have also failed to confirm the findings of Schneider and Sherman (Banker, Hunt, & Pagano, 1969; Jamieson & Albert, 1970; Gold & King, 1971), but if these findings should turn out to be viable it would be necessary to distinguish the possible influences of FS on memory storage processes from the influences of FS on the effectiveness of an ECS treatment. The examination of brain seizure threshold and the characteristics of seizure activity with and without prior FS may serve to distinguish between these two possibilities.

Recently Misanin, Miller, and Lewis (1968) proposed that well-established memories may become susceptible to ECS at the time they are retrieved or "reinstated." In support of this view, they reported that RA was produced even 1 day after training if a stimulus previously associated with punishment was presented just prior to the ECS treatment. However, in a recent attempt to replicate and extend these findings, we failed completely to obtain the "reinstatement" susceptibility effect (Dawson & McGaugh, 1969). Despite these inconsistencies in experimental findings, the potential importance of posttrial stimulus events in studies of memory storage cannot be overlooked. The bases of the discrepancies in findings will require additional research.

VII. Relationships among Memory Storage Processes

The problem of the relationships among memory storage processes is emerging as perhaps the most complex issue in memory storage research and the answers to the questions raised may turn out to be the most important contributions in this general area of research. Until recently it was thought that agents affecting memory storage do so by interfering with some labile phase of memory which was essential for the formation of a consolidated trace. Retrieval was also generally assumed to occur only from this consolidating trace, following the notions proposed by Hebb (1949). However, studies using antibiotics have shown that when the drugs are given prior to training, "learning" may be unaffected but retention declines within hours or days (Agranoff, Chapter 6; Barondes, 1968). With posttrial injections amnesia is not produced at the time of treatment, but rather over a period of time (Agranoff, Davis, Lim, & Casola, 1968). The major question is whether such evidence indicates that the STM trace is temporary and uninfluenced by amnesic agents or whether amnesic agents simply act by potentiating forgetting (Deutsch, 1969).

A prior word of caution is in order. In general, these studies using antibiotics (Barondes, 1968) have employed minimal training. When treatments are given prior to the initial training, it is difficult to eliminate the possibility that the treatments affect acquisition. If this is the case, then differences in retention between groups might simply reflect initial acquisition differences (cf. Underwood, 1964). This is, of course, not a serious problem when posttrial treatments are given since groups may be equated for initial level of acquisition, that is, after training but prior to treatment.

Short-term decay of memory has also been found using posttrial ECS treatments (Geller & Jarvik, 1968b). Typically, RA following ECS is assessed 24 hours after training, and under these circumstances animals trained on an inhibitory avoidance task and given ECS shortly after training exhibit almost total amnesia. However, somewhat surprisingly, the animals show excellent retention of the avoidance response if they are tested an hour following the ECS treatment. If tested 6 hours after the ECS they show almost complete amnesia. The amnesia appears to develop over a period of several hours following the ECS treatment. McGaugh and Landfield (1970) and McGaugh, Landfield and Dawson (in preparation) have recently shown that such short-term retention occurs if the training-ECS interval is 20 seconds, but not if ECS is given 5 seconds after training (see Fig. 9). Moreover, it is clear that the effect is not due simply to some nonspecific effect of a FS or ECS treatment. Controls which are

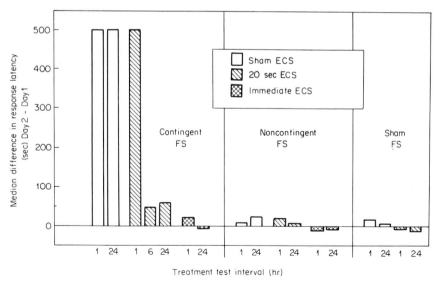

Fig. 9. Short term decline of retention of a one-trial inhibitory task avoidance following an ECS. The experimental groups given an ECS 20 seconds following the trial show virtually complete retention 1 hour later yet little retention at 6 hours or 24 hours. No comparable effect is seen in groups given only ECS, or noncontingent footshock or both treatments. (From McGaugh & Landfield, 1970.)

untrained but which are given FS and ECS or ECS alone do not show a comparable effect when tested 1 hour after such treatments. Such findings have been interpreted (Deutsch, 1969) as evidence that ECS may act to merely hasten forgetting rather than to prevent storage. Such an interpretation is difficult to accept for the following reasons: First, a 5-second ECS prevents the appearance of short-term retention, whereas a 20-second ECS does not. In subsequent studies of this problem shorter training–retention intervals should be employed to assure that retention is not possible for a brief period following training (i.e., less than 1 hour). Second, the degree of forgetting eventually attained decreases as the time between the training and the amnesic treatment increases; all animals do not reach the same level of amnesia. Obviously, a somewhat more complex hypothesis is required than the assumption that ECS merely causes memory decay.

In addition to this evidence, several related findings also suggest that multiple processes might be involved in memory storage (cf. McGaugh, 1968a). Recent findings suggest that it is possible to restart storage processes by giving additional treatments, but only if the treatments are given within a few hours following training. Such effects have been obtained with anodal brain stimulation given after cathodal stimulation, strychnine given after ECS, and amphetamine given after a protein synthesis inhibitor (Albert, 1966; Barondes & Cohen, 1968; McGaugh, 1968b). In all cases the effect was graded—the effectiveness decreased with time of administration of the second treatment. It would appear to be very difficult to interpret such effects without assuming at least two memory processes, STM and LTM.

In summary, although the theoretical speculations have developed from the simple notion of a labile phase of memory producing a durable long-term trace, the evidence seems to require a much more complex picture of memory storage. At this point, we wish to introduce several models which attempt to account for some of the phenomena of RA discussed above. These models are not intended to be exhaustive. They should serve rather as a basis from which questions may be asked. We hope that the clear differences in predictions from several of these models may be useful.

VIII. A Storage and a Retrieval Effect of ECS

Figure 10 shows the model recently proposed by Weiskrantz (1966). According to this model STM and LTM are based on different processes. A training trial produces an STM trace which after a few seconds "primes" or initiates the growth of LTM. An ECS given within a few seconds disrupts STM thereby preventing the priming of LTM. An ECS given after the priming occurs does not affect the development of LTM but, by generating "neural noise," prevents the retrieval of LTM until the noise subsides. X and Y indicate the

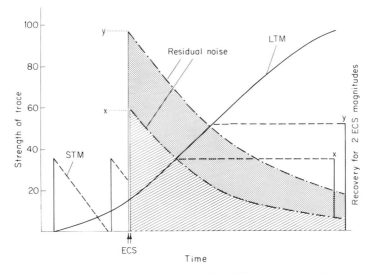

Fig. 10. A set of possible relationships among STM, LTM, and noise. LTM is assumed to require a short priming period by STM before it becomes viable. ECS delivered during this period prevents such priming. ECS delivered at any time produced noise. X and Y represent two different degrees of severity of ECS treatment, giving rise to two different levels of noise. (Based on model proposed by Weiskrantz, 1966.)

amount of "noise" generated by low- and high-intensity ECS. Recovery from amnesia occurs within a shorter interval following the low-intensity ECS (X). If ECS is given after completion of the short "priming period" full recovery occurs when its "noise" level becomes lower than the stimulus trace strength. Thus, permanent RA is produced with short-training trial-ECS intervals and temporary amnesia is produced with longer intervals.

The hypothesis successfully accounts for a number of phenomena, including brief RA gradients (e.g., Chorover & Schiller, 1965, 1966), recovery from RA with low intensity ECS (when obtained, e.g., Pagano *et al.,* 1969), and the effects of variations in the effectiveness of different amnesic treatments (e.g., Alpern & McGaugh, 1968; Dorfman & Jarvik, 1968a). As presented, the model fails to account for several phenomena including (1) the permanence of RA (e.g., Luttges & McGaugh, 1967; Zornetzer & McGaugh, 1969), (2) the STM observed after amnesic treatments (e.g., Geller & Jarvik, 1968b; McGaugh & Landfield, 1970), (3) the long gradients of RA and retrograde facilitation, which were analyzed in some detail above, and (4) the lack of effect of ECS on learned responses when given prior to retention tests (e.g., McGaugh & Longacre, 1969; Zerbolio, 1969).

Figure 11 illustrates a modified version of a model proposed by Cherkin (1969). According to this model, weak amnesic treatments retard but do not stop consolidation. The differential amnesic effects of different treatments are assumed to be due to their variable effectiveness in retarding consolidation. The figure shows the way in which consolidation rate might be altered by one level of amnesic treatment. The RA gradient is graphed by plotting above each ECS treatment the point indicating the maximum strength of trace attained by the curve generated by that treatment. A more effective treatment would produce

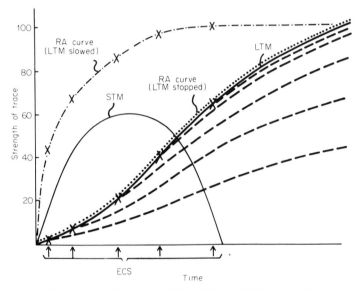

Fig. 11. Possible relationships among STM, LTM, and RA curves. STM and LTM are assumed to be independent and ECS effects only LTM by slowing consolidation. If performance is assessed some time after the treatment is given, consolidation will have proceeded to a higher level. Because of this the RA curve will be measured to asymptote more rapidly than the curve of LTM consolidation. (Modified from Cherkin, 1969.)

greater slowing (i.e., flatten the slopes) and thus generate a "long" gradient of RA. This model assumes that STM is unaffected by ECS and predicts that short-term retention curves all asymptote at the same point in time, regardless of the training–ECS interval (see Fig. 11). As presented, it is not clear from this model why different rates of consolidation should result in different asymptotes. If consolidation is only retarded, eventually all traces should reach the same asymptote. This would suggest that all RA should be but temporary.

A modification of Cherkin's model is shown in Fig. 12. The RA curve is plotted as in Fig. 11. This model assumes that STM is essential for LTM. ECS impairs consolidation by speeding the decline of STM. The asymptote of LTM is determined by the duration of STM. Thus STM and LTM are linked and STM is impaired by ECS. This model, therefore, predicts that as the treatment strength and/or delay is varied the degree of both STM and LTM will change. Decay of STM in this figure is depicted by the thin dashed lines which descend from the STM curve at the point where ECS is indicated. This modified model appears to account for a number of phenomena including (1) STM decay, (2) variations in RA gradients, (3) the permanence of RA, and (4) lack of, or decrease in STM with immediate posttrial ECS.

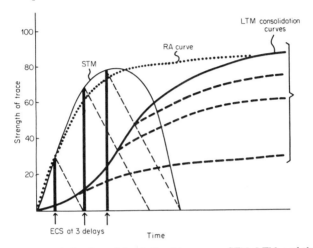

Fig. 12. Further analysis of possible relationships among STM, LTM, and the RA curves. STM is assumed to be essential for LTM. ECS is assumed to slow consolidation by speeding the decay of STM.

The major difference between the models depicted in Figs. 11 and 12 is in relation to the effects of memory-impairing treatments upon STM. If ECS and memory-impairing drugs affect STM by causing it to decay or stop, STM gradients will reach a maximum at points which are a function of the magnitude and/or delay of the treatment (see Fig. 12). If STM is not affected, however, STM gradients will all reach a maximum at the same point regardless of the treatment parameters. If ECS causes LTM consolidation to stop, then this gradient can be assumed to be a good estimate of the duration of consolidation If, on the other hand, ECS only slows consolidation, then an underestimation of consolidation duration (as measured after 24 hours) is likely. This latter assumption also predicts some slight recovery of memory at some time after 6 hours when STM has decayed. If STM and LTM are linked, and ECS affects STM, then the duration of STM should bear some relationship to the degree of amnesia in LTM.

Of the several models considered here, the one shown in Fig. 12 appears to account best for the bulk of the relevant data from studies using ECS. There are many implications of each model however, and the development of a completely adequate set of hypotheses will require a considerable amount of additional research. Furthermore, the adequacy of each in handling the phenomena of RA obtained with other amnesic agents will need to be assessed. Some findings, such as the delayed "decay" of STM following antibiotics (Agranoff, Chapter 6) do not seem to fit any of the simplified models presented here. Obviously, it is essential that we have a clear understanding of the basic phenomena in this area of research if we are eventually to understand their neurobiological bases.

IX. Neurobiological Bases of Treatments

The classic assumptions have been that memory storage consisted of a labile phase, presumably electrical in nature, leading to a permanent structural phase, presumably biochemical in nature. These assumptions were supported primarily by the evidence that treatments such as ECS disrupted the labile electrical phase but had no disruptive effect on LTM. The above discussions suggest that the situation is much more complex. The findings suggesting that STM may persist for a short time following an amnesic treatment indicate that the interplay between electrophysiological and biochemical processes involved in memory is quite complex.

As suggested by John (1967), it may be the persistence or reoccurrence of electrical patterns which is the necessary condition for memory retrieval. Thus, in the early stages of memory storage, disruption might be achieved not by directly eliminating storage-specific neural firing, but perhaps by depleting brain energy reserves and transmitter substances. As a result, the biochemical consequences of patterned firing would be reduced. Short-term decay could simply reflect the disappearance of the reserves. If this assumption is valid, then we must further suggest that attention should be directed toward the electrophysiological specificity during storage. It may very well be that the biochemical consequences in individual neurons are *not* as specific. Moreover, the same kinds of changes may be taking place in a great many neurons in restricted regions of the brain. Evidence from studies on labeled precursor uptake into RNA during learning (Zemp, Wilson, & Glassman, 1967) is promising. These experiments, as well as those employing antibiotic drugs (e.g., Agranoff, Chapter 6; Barondes, 1968), provide strong support for the hypothesis that memory storage involves macromolecular synthesis.

As yet, few clues to storage mechanisms have been provided by neurophysiological studies. But hopefully, an understanding of the correlations between electrical activity of the brain and memory impairment and facilitation may prove to be of help (Zornetzer & McGaugh, 1971). A recent electrophysiological study by Luttges (1968) has provided some evidence relevant to the analysis of facilitative effects of drugs which bears directly upon a question which has been the source of some concern. Does the fact that a wide variety of drugs have been shown to facilitate learning indicate that many different processes are involved, or is there some commonality of drug effects?

Luttges (1968) investigated two drugs, megimide and strychnine, which had been shown to exert both facilitative and disruptive effects upon memory storage depending upon the drug dose used, but which had different central mechanisms of action. He argued that effects common to both drugs injected in facilitatory doses might be used to assess a common facilitatory mechanism. Effects common to both drugs injected in disruptive doses might likewise be used to assess a common disruptive mechanism. Effects common to both facilitative and disruptive doses might, he argued, be assumed to be incidental to the main effect. In order to examine this possibility Luttges studied the effects of intraperitoneal injections of these drugs on spontaneous and evoked activity in a number of cortical and subcortical areas in chronically prepared unanesthetized rats. Recovery cycle data and central brain stem stimulation were used in conjunction with ongoing neurophysiological recording as indicators of responsivity of peripheral and central brain regions.

Summarized briefly, the results reveal a minimal effect of both drugs in the lower sensory relays. Strychnine effects appear greatest as changes in evoked and spontaneous activity in the reticular formation, with some changes in the posterior hypothalamus. Megimide injections appear to cause changes in the posterior hypothalamus mainly and in the cortex to a lesser extent. Despite the apparent differences in the systems which appear to change following injection of the two drugs, there is some degree of commonality between the two drugs. With facilitative doses of both drugs, the posterior hypothalamus appeared to be activated. However, the hippocampus and reticular formation revealed more sustained and slower activity than with disruptive doses. With disruptive doses, the situation was somewhat the reverse. Hippocampus and reticular formation appeared to be activated along with the posterior hypothalamus.

On the basis of his findings Luttges suggested that with facilitative doses of both drugs high states of activation may be produced without the need for great amounts of stimulus input, thus decreasing the possibility of retroactive interference. Thus, there is evidence for a close tie between what were previously suggested as two distinct hypotheses for facilitative effects of drugs. The first is the notion of direct enhancement of storage processes and the second the possibility of reduced interference—here they appear to be inextricably related. Studies by Nasello and Izquierdo (1969) also appear to implicate the hippocampus as a structure involved in drug facilitation of memory storage.

Studies of brain stimulation and lesions have also provided some evidence concerning anatomical structures which may be critically involved in memory storage. Facilitation of memory storage has been obtained with electrical and chemical stimulation of the mesencephalic reticular formation (Alpern, 1968;

Bloch, Denti, & Schmaltz, 1966; Denti, McGaugh, Landfield, & Shinkman, 1970) and hippocampus (Erickson & Patel, 1969; Stein & Chorover, 1968). Impairment of memory storage has been obtained with electrical stimulation, lesions, and spreading depression in cortical and limbic structures and basal ganglia (e.g., Avis & Carlton, 1968; Dorfman, Bohdanecka, Bohdanecky, & Jarvik, 1969; Hudspeth, 1969; Kesner & Doty, 1968; Wyers, Peeke, Williston, & Herz, 1968). These recent findings clearly suggest that it may be possible to discover the neural systems involved in memory storage.

Research on the problem of memory storage has not, as yet, provided a clear and detailed understanding of either the structures or the processes involved in recording experiences, nor is it likely to do this in the near future. Nonetheless, progress is being made and such understanding is essential for a comprehensive theory of animal theory.

REFERENCES

Agranoff, B. W., Davis, R. E., Lim, R., & Casola, L. Biological effects of antimetabolites used in behavioral studies. In D. H. Efron (Ed.), *Psychopharmacology: A review of progress*. Washington, D.C.: US Govt. Printing Office, 1968. PHS Publ. No. 1836, pp. 909–917.

Albert, D. J. The effects of polarizing currents on the consolidation of learning. *Neuropsychologia*, 1966, **4**, 65–77.

Alpern, H. P. Facilitation of learning by implantation of strychnine sulfate in the central nervous system. Unpublished doctoral dissertation, University of California, Irvine, 1968.

Alpern, H. P., & McGaugh, J. L. Retrograde amnesia as a function of duration of electroshock stimulation. *Journal of Comparative and Physiological Psychology*, 1968, **65**, 265–269.

Avis, H. H., & Carlton, P. L. Retrograde amnesia produced by hippocampal spreading depression. *Science*, 1968, **161**, 73–75.

Banker, G., Hunt, E., & Pagano, R. Evidence supporting the memory disruption hypothesis of electroconvulsive shock action. *Physiology & Behavior*, 1969, **4**, 895–899.

Barondes, S. H. Effect of inhibitors of cerebral protein synthesis on "long-term" memory in mice. In D. H. Efron (Ed.), *Psychopharmacology: A review of progress*. Washington, D.C.: US Govt. Printing Office, 1968. PHS Publ. No. 1836, pp. 905–908.

Barondes, S. H., & Cohen, H. D. Arousal and the conversion of "short-term" to "long term" memory. *Proceedings of The National Academy of Sciences of The United States*, 1968, **61**, 923–929.

Bloch, V., Denti, A., & Schmaltz, G. Effets de la stimulation reticulaire sur la phase de consolidation de la trace amnésique. *Journal de Physiologie (Paris)*, 1966, **58**, 469–470.

Cherkin, A. Kinetics of memory consolidation: Role of amnesic treatment parameters. *Proceedings of The National Academy of Sciences of The United States*, 1969, **63**, 1094–1101.

Chevalier, J. A. Permanence of amnesia after a single posttrial electroconvulsive seizure. *Journal of Comparative and Physiological Psychology*, 1965, **59**, 125–127.

Chorover, S. L., & Schiller, P. H. Short term retrograde amnesia in rats. *Journal of Comparative and Physiological Psychology*, 1965, **59**, 73–78.

Chorover, S. L., & Schiller, P. H. Re-examination of prolonged retrograde amnesia in one-trial learning. *Journal of Comparative and Physiological Psychology*, 1966, **61**, 34–41.

Davis, R. E., & Agranoff, B. W. Stages of memory formation in goldfish: Evidence for an environmental trigger. *Proceedings of The National Academy of Sciences of The United States*, 1966, **55**, 555–559.

Dawson, R. G. Retrograde amnesia and CER incubation re-examined. *Psychological Bulletin*, 1971, **75**, 278–285.

Dawson, R. G., & McGaugh, J. L. Electroconvulsive shock effects on a reactivated memory trace: Further examination. *Science*, 1969, **166**, 525–527.

Denti, A., McGaugh, J. L., Landfield, P., & Shinkman, P. Facilitation of learning with post-trial stimulation of the reticular formation. *Physiology & Behavior*, 1970, **5**, 659–662.

Deutsch, J. A. The physiological basis of memory. *Annual Review of Psychology*, 1969, **20**, 85–104.

Dingman, W., & Sporn, M. B. Molecular theories of memory. *Science*, 1964, **144**, 26–29.

Dorfman, L. J., Bohdanecka, M., Bohdanecky, Z., & Jarvik, M. E. Retrograde amnesia produced by small cortical stab wounds in the mouse. *Journal of Comparative and Physiological Psychology*, 1969, **69**, 324–328.

Dorfman, L. J., & Jarvik, M. E. A parametric study of electroshock induced retrograde amnesia in mice. *Neuropsychologia*, 1968, **6**, 373–380. (a)

Dorfman, L. J., & Jarvik, M. E. Comparative amnesic effects of transcorneal and transpinnate ECS in mice. *Physiology & Behavior*, 1968, **3**, 815–818. (b)

Duncan, C. P. The retroactive effect of electroshock on learning. *Journal of Comparative Psychology*, 1949, **42**, 32–44.

Erickson, C. K., & Patel, J. B. Facilitation of avoidance learning by post-trial hippocampal electrical stimulation. *Journal of Comparative and Physiological Psychology*, 1969, **68**, 400–406.

Fishbein, W., McGaugh, J. L., & Swarz, J. R. Retrograde amnesia: Electroconvulsive shock effects after termination of rapid eye movement sleep deprivation. *Science*, 1971, **172**, 80–82.

Franchina, J. T., & Moore, M. H. Strychnine and the inhibition of previous performance. *Science*, 1968, **160**, 903–904.

Garg, M., & Holland, H. C. Consolidation and maze learning: The effects of post-trial injections of a stimulant drug (Picrotoxin). *Psychopharmacologia*, 1968, **40**, 15–16.

Geller, A., & Jarvik, M. E. Electroconvulsive shock induced amnesia and recovery. *Psychonomic Science*, 1968, **10**, 15–16. (a)

Geller, A., & Jarvik, M. E. The time relations of ECS-induced amnesia. *Psychonomic Science*, 1968, **12**, 169–170. (b)

Gerard, R. W. Physiology and psychiatry. *American Journal of Psychiatry*, 1949, **106**, 161–173.

Glassman, E. The biochemistry of learning: An evaluation of the role of RNA and protein. *Annual Review of Biochemistry*, 1969, **38**, 605–616.

Glickman, S. E. Perseverative neural processes and consolidation of the memory trace. *Psychological Bulletin*, 1961, **58**, 218–233.

Gold, P. E., & King, R. E. Proactive effects of footshock-ECS pairing: Failures to obtain the effect. In preparation.

Greenough, W. T., Schwitzgebel, R. L., & Fulcher, J. K. Permanence of ECS-produced amnesia as a function of test conditions. *Journal of Comparative and Physiological Psychology*, 1968, **66**, 554–556.

Hebb, D. O. *The organization of behavior.* New York: Wiley, 1949.

Herz, M. J., & Peeke, H. V. S. ECS produced retrograde amnesia: Permanence vs. recovery over repeated testing. *Physiology & Behavior*, 1968, **3**, 517-521.

Hudspeth, W. J. Retrograde amnesia: Time dependent effects of rhinencephalic lesions. *Journal of Neurobiology*, 1969, **2**, 221-232.

Jamieson, J. L., & Albert, D. J. Amnesia from ECS: The effect of pairing ECS and footshock. *Psychonomic Science*, 1970, **18**, 14-15.

John, E. R. *Mechanisms of memory*. New York: Academic Press, 1967.

Kesner, R. P., & Doty, R. W. Amnesia produced in cats by local seizure activity initiated from the amygdala. *Experimental Neurology*, 1968, **21**, 58-68.

Krivanek, J., & McGaugh, J. L. Effects of pentylenetetrazol on memory storage in mice. *Psychopharmacologia*, 1968, **12**, 303-321.

Krivanek, J., & McGaugh, J.L. Facilitating effects of pre- and posttrial amphetamine administration on discrimination learning in mice. *Agents and Actions*, 1969, **1**, 36-42.

Lajtha, A., & Toth, J. Instability of cerebral proteins. *Biochemical and Biophysical Research Communications*, 1966, **23**, No. 3, 294-298.

Luttges, M. W. Electrophysiological dose-response effects of megimide and strychnine. Unpublished doctoral dissertation. University of California, Irvine, 1968.

Luttges, M. W., & McGaugh, J. L. Permanence of retrograde amnesia produced by electroconvulsive shock. *Science*, 1967, **156**, 408-410.

McGaugh, J. L. Time-dependent processes in memory storage. *Science*, 1966, **153**, 1351-1358.

McGaugh, J. L. A multi-trace view of memory storage processes. In D. Bovet, N. H. F. Bovet, and A. Oliverio (Eds.), Attuali orientamenti della ricerca sull'apprendimento e la memoria. *Academia Nazionale dei Lincei*, 1968, **109**, 13-24.(a)

McGaugh, J. L. Drug facilitation of memory and learning. In D. H. Efron (Ed.), *Psychopharmacology: A review of progress.* Washington, D.C.: US Govt. Printing Office, 1968. PHS Publ. No. 1836, pp. 891-904.(b)

McGaugh, J. L. Facilitation of memory storage processes. In S. Bogoch (Ed.), *The future of the brain sciences.* New York: Plenum Press, 1969. Pp. 355-370.

McGaugh, J. L., & Alpern, H. P. Effects of electroshock on memory: Amnesia without convulsions. *Science*, 1966, **152**, 665-666.

McGaugh, J. L., & Herz, M. J. *Memory consolidation.* San Francisco: Albion Publishing Co., 1971, in press.

McGaugh, J. L., & Krivanek, J. Strychnine effects on discrimination learning in mice: Effects of dose and time of administration. *Physiology & Behavior,* 1970, **5**, 1437-1442.

McGaugh, J. L., & Landfield, P. Delayed development of amnesia following electroconvulsive shock. *Physiology & Behavior,* 1970, **5**, 1109-1113.

McGaugh, J. L., & Longacre, B. Effect of electroconvulsive shock on performance of a well-learned avoidance response: Contribution of the convulsion. *Commun. Behav. Biol.,* 1969, **4**, 177-181.

McGaugh, J. L., & Petrinovich, L. F. Effects of drugs on learning and memory. *International Review of Neurobiology,* 1965, **8**, 139-196.

McGaugh, J. L., Alpern, H. P., & Luttges, M. W. Further analysis of retrograde amnesia as a function of the footshock-electroconvulsive shock interval. In preparation.

McGaugh, J. L., Landfield, P. W., & Dawson, R. G. Delayed development of amnesia following electroconvulsive shock: Further analysis. In preparation.

Miller, A. J. Variations in retrograde amnesia with parameters of electroconvulsive shock and time of testing. *Journal of Comparative and Physiological Psychology,* 1968, **66**, 40-47.

Misanin, J. R., Miller, R. R., & Lewis, D. J. Retrograde amnesia produced by electroconvulsive shock after reactivation of a consolidated memory trace. *Science,* 1968, **160,** 554–555.

Nasello, A. G., & Izquierdo, I. Effect of learning and of drugs on the ribonucleic acid concentration of brain structures of the rat. *Experimental Neurology,* 1969, **23,** 521–528.

Nielson, H. C. Evidence that electroconvulsive shock alters memory retrieval rather than memory consolidation. *Experimental Neurology,* 1968, **20,** 3–20.

Oliverio, A. Neurohumoral systems and learning. In D. H. Efron, (Ed.), *Psychopharmacology: A review of progress.* Washington, D.C.: US Govt. Printing Office, 1968. PHS Publ. No. 1836, pp. 891–904.

Pagano, R. D., Bush, D. F., Martin, G., & Hunt, E. B. Duration of retrograde amnesia as a function of electroconvulsive shock intensity. *Physiology & Behavior,* 1969, **4,** 19–21.

Pinel, J. P. J. A short gradient of ECS-produced amnesia in an appetitive situation. *Journal of Comparative and Physiological Psychology,* 1969, **68,** 650–655.

Pinel, J. P. J., & Cooper, R. M. The relationship between incubation and ECS gradient effects. *Psychonomic Science,* 1966, **6,** 125–126.

Ray, O. S., & Barrett, R. J. Disruptive effects of electroconvulsive shock as a function of current level and mode of delivery. *Journal of Comparative and Physiological Psychology,* 1969, **67,** 110–116.

Reichert, H. Unpublished doctoral dissertation, University of Waterloo, Waterloo, Ontario, Canada, 1968.

Robustelli, F., Geller, A., & Jarvik, M. E. Detention, electroconvulsive shock, and amnesia. *Proceedings of The 76th Annual Convention, American Psychological Association, 1968,* Pp. 331–332.

Schneider, A. M., Kapp, B., Aron, C., & Jarvik, M. E. Retroactive effects of transcorneal and transpinnate ECS on step-through latencies in mice and rats. *Journal of Comparative and Physiological Psychology,* 1969, **69,** 506–509.

Schneider, A. M., & Sherman, W. Amnesia: A function of the temporal relation of footshock to electroconvulsive shock. *Science,* 1968, **59,** 219–221.

Spevack, A. A., & Suboski, M. D. Retrograde effects of electroconvulsive shock on learned responses. *Psychological Bulletin,* 1969, **72,** 66–76.

Stein, D. G., & Chorover, S. L. Effects of posttrial electrical stimulation of hippocampus and caudate nucleus on maze learning in the rat. *Physiology & Behavior,* 1968, **3,** 787–791.

Stephens, G., & McGaugh, J. L. Retrograde amnesia effects of periodicity and degree of training. *Communications in Behavioral Biology (A),* 1968, **1,** 267–275. (a)

Stephens, G. J., & McGaugh, J. L. Periodicity and memory in mice: A supplementary report. *Communications in Behavioral Biology (A),* 1968, **2,** 59–63. (b)

Stephens, G. J., McGaugh, J. L., & Alpern, H. P. Periodicity and memory in mice. *Psychonomic Science,* 1967, **8,** 201–202.

Stratton, L. O., & Petrinovich, L. F. Post-trial injections of an anti-cholinesterase drug and maze learning in two strains of rats. *Psychopharmacology,* 1963, **5,** 47–54.

Tenen, S. S. Retrograde amnesia from electroconvulsive shock in a one-trial appetitive learning task. *Science,* 1965, **148,** 1248–1250. (a)

Tenen, S. S. Retrograde amnesia. *Science,* 1965, **149,** 1521. (b)

Thompson, R. C., & Ballou, J. E. Studies of metabolic turnover with tritium as a tracer. V. The predominantly non-dynamic state of body constituents of the rat. *Journal of Biological Chemistry,* 1956, **223,** 795–809.

Underwood, B. J. Degree of learning and the measurement of forgetting. *Journal of Verbal Learning and Verbal Behavior,* 1964, **3,** 112–129.

von Hungen, K., Mahler, H. R., & Moore, W. J. Turnover of protein and ribonucleic acid in synaptic subcellular fractions from rat brain. *Journal of Biological Chemistry,* 1968, **243,** 1415–1423.

Weiskrantz, L. Experimental studies of amnesia. In C. W. M. Whitty and O. L. Zangwill (Eds.), *Amnesia.* London and Washington, D.C.: Butterworths, 1966. Pp. 1–35.

Westbrook, W. H., & McGaugh, J. L. Drug facilitation of latent learning. *Psychopharmacologia,* 1964, **5,** 440–446.

Wyers, E. J., Peeke, H. V. S., Williston, J. S., & Herz, M. J. Retroactive impairment of passive avoidance learning by stimulation of the caudate nucleus. *Experimental Neurology,* 1968, **22,** 350–366.

Zemp, J. W., Wilson, J. E., & Glassman, E. Brain function and macromolecules. II. Site of increased labelling of RNA in brains of mice during a short-term training experience. *Proceedings of The National Academy of Sciences of The United States,* 1967, **58,** No. 3, 1120–1125.

Zerbolio, D. J. The proactive effect of electroconvulsive shock on memory storage: With and without convulsions. *Communications in Behavioral Biology (A),* 1969, **4,** 23–27.

Zinkin, S., & Miller, A. J. Recovery of memory after amnesia induced by electroconvulsive shock. *Science,* 1967, **155,** 102–104.

Zornetzer, S., & McGaugh, J. L. Effects of electroconvulsive shock upon inhibitory avoidance: The persistence and stability of amnesia. *Communications in Behavioral Biology (A),* 1969, **3,** 173–180.

Zornetzer, S., & McGaugh, J. L. Effects of frontal brain electroshock stimulation on EEG activity and memory in rats: Relationships to ECS-produced retrograde amnesia. *Journal of Neurobiology,* 1970, **1,** 379–394.

Chapter 6

EFFECTS OF ANTIBIOTICS ON LONG-TERM MEMORY FORMATION IN THE GOLDFISH[1]

Bernard W. Agranoff

I. Introduction

The use of disruptive agents such as trauma, electroconvulsive shock, and convulsant drugs as experimental probes in improving our understanding of brain memory mechanisms is well known. A process of consolidation or fixation of memory is inferred from the fact that a learned task becomes decreasingly susceptible to disruption as a function of time following learning. Varying effects of disruption are observed upon retraining as a function of the original training task, the kind of experimental subject, and the nature of the disruptive agent. Gradients varying from a few seconds to many hours in duration are derived, representing the growing resistance of a new memory to disruption during the posttraining period. It is immediately apparent that in order to explain the many time courses of consolidation which have been reported, one must either assume that there are many stages of consolidation—the disruption of any of which can modify an animal's ability to perform a learned behavior—or that each agent artifactually interferes with memory in a way that is more characteristic of its mode of action than descriptive of a physiological process involved in memory formation. Since all of the above agents temporarily interrupt gross brain function by producing convulsions and/or unconsciousness, a simple explanation of their action involves the dissolution of "reverberating" electrical circuits within the neuronal network which ultimately would have become converted to some more stable form without the interruption.

That memory in its permanent form should be chemical is suggested by a number of considerations. The long-lasting nature of memory suggests that it

[1] The research reported in this chapter was supported by grants from the National Institute of Mental Health and the National Science Foundation. The able assistance of Mr. Paul Klinger is gratefully acknowledged. The findings reported here were also presented at the "Institute de la Vie" in Versailles in July 1969.

ultimately resides in a form resistant to various natural forces tending to randomize molecular arrangements, i.e., in the form of covalent chemical bonds. Such bonds are known to mediate the storage of genetic material. The DNA that contains coded instructions for the synthesis of specific enzymes resists physical agents as well as the enzymes that act on the rapidly turning over intracellular components engaged in energy production for cellular function. More direct evidence for a molecular basis of memory stems from experiments made possible by the recent availability of antibiotic agents that block macromolecular processes in animal cells. These novel substances, to be considered in some detail in this chapter, have been shown to block memory formation when they are applied to the brain. Unlike previously used amnesic agents described above, they do not produce gross neurological disturbance.

II. The Antibiotic Inhibitors of Macromolecular Synthesis

The molecular business of a cell might conveniently be divided into two categories: those related to growth (including cellular reproduction) and those related to energetics (the conversion of food to useful energy). The machinery for reproduction and growth is located principally within the nucleus of the animal cell, while energy metabolism goes on in various extranuclear structures. The interrelationship of these two functions of cells is complex. The nucleus regulates the level of key metabolic enzymes, and thus plays a role in cellular energetics as well as growth. The macromolecular synthesis of growth can on the other hand not occur without the presence of high energy substances provided by the cytoplasm, such as ATP. Despite this interdependence, specific chemical agents exist that can selectively block a given step of either growth or metabolism. The effects of such agents may be analyzed to give further insight into these two domains of cellular physiology.

Disruption of energy metabolism of the brain of higher animals by such substances as cyanide or carbon monoxide often produces irreversible effects and does not appear to be a useful means of understanding brain function. Agents that block macromolecular synthesis are relatively new chemical tools. They were first discovered and explored for their possible use as antiinfectious and anticancer agents. In nature, they presumably arose via natural selection and mutation and provide an armamentarium for microorganisms and some higher life forms. Their secretion by a microorganism into its growth medium results in the inhibition of growth of neighboring organisms. The specific molecular mechanism by which such metabolic monkey wrenches work varies, but in general each agent has a single selective molecular site at which its acts. These agents are therefore useful experimental probes. Of particular interest are those agents that block DNA, RNA, or protein synthesis.

Since the mechanism of the synthesis of DNA, RNA, and protein is very similar in all living material, it is not surprising that many of these antibiotic blocking agents of macromolecular synthesis are biologically active in animal as well as in bacterial systems. For example, arabinosyl cytosine selectively blocks DNA synthesis (replication) in bacteria as well as in higher animals. Actinomycin D has no effect on replication, but by combining it with DNA its use as a template for RNA synthesis (transcription) is prevented. These activities are confined to the nucleus of the eukaryotic cell. Translation, the synthesis of protein by ribosomes, although a part of the cellular growth process, occurs mainly in the cytoplasm of the cell. It may be blocked by a number of agents including puromycin and the cycloheximides. Proteins are the final effectors and express their function via their catalytic (enzymic) activities. The extranuclear site in which most proteins are made is the polysome, which may be considered a factory in which a coded messenger RNA (mRNA) is sequentially read as it passes down the assembly line, by a series of workmen (the ribosomes), each of whom assembles amino acids into an identical continuous sequence determined by the mRNA. The existence of 20 known amino acids which can be coded and the usual length of proteins (100+ amino acids) make possible the assembly of more unique protein molecules than could be assembled from the atoms available on this planet! In the polysome, mRNA determines which combination of amino acids will be put together, and in a given polysome the same mRNA can give rise to as many molecules of a given protein as there are ribosomes. The antibiotic puromycin interferes with this process by attaching itself to the growing peptide chain. The peptidyl puromycin formed leaves the polysome. This antibiotic thus chops off incomplete fragments of protein. The glutarimide inhibitors, cycloheximide (CXM) and acetoxycycloheximide (AXM),[2] slow down the assembly process. They therefore also block formation of peptidyl puromycin. Both puromycin and the cycloheximides have the net effect of preventing the formation of proteins and hence cellular growth as well. They also block some forms of cellular regulation, such as enzyme induction. Although they act by different mechanisms, the net effect is the inhibition of protein synthesis.

The neuron, like other cells, synthesizes RNA and protein. The polysomes are near the nucleus, in the perikaryon, but active neuronal metabolism may be occurring at a distance, in presynaptic terminals separated from the rest of the cell via an axon which might be as little as a few microns or as much as a meter away. Regulation by the nucleus may require transport of macromolecules via the axon.

---

[2] The data for AXM are provided in Agranoff et al. (1966).

III. Effects of Antibiotics on Learning and Memory in the Goldfish

A. DESCRIPTION OF TECHNIQUE

Several years ago, we developed a technique for the intracranial injection of various materials in a 5–10 μl volume into unanesthetized goldfish (Agranoff & Klinger, 1964). By means of isotopic studies we found that injections of antibiotics such as puromycin (170 μg) or AXM (0.2 μg) inhibited normal protein synthesis throughout the brain by about 80% for several hours, with recovery in about 1 day (Brink, Davis, & Agranoff, 1966; Lim, Brink, & Agranoff, 1970). Despite this drastic effect on protein synthesis, there was little change in the gross behavior of the animals. In fact, injected goldfish could be trained in a shock-avoidance task with no gross evidence of impairment in performance (Agranoff, Davis, & Brink, 1965). The apparatus we used in our initial studies (Task 1) was modeled after a two-way shuttle box of Bitterman (Behrend & Bitterman, 1964). The tasks differed mainly in that our fish were given a discrete number of trials rather than trained to a criterion. Animals were given Trials 1–20 on Day 1 of an experiment, were returned to home tanks, then were given 10 retraining trials (Trials 21–30) 3 days later (Day 4). Between training sessions fish were stored in individual plastic home tanks and were maintained without feeding in continuous illumination. Possible influences of daily rhythms on behavior or on drug effects have not been investigated systematically, but there was no indication that the time of day at which animals were trained influenced the experimental results. One-minute trials were given in blocks of five, and the blocks alternated with 5 minutes of darkness in the shuttle box. At the beginning of a trial, a light (Sylvania 120PSB) went on on the side of the shuttle box in which the fish had been placed. Twenty seconds later, intermittent electrical shock (0.2 seconds, off time 1.3 seconds; 3 V; 60 cps at 0.1 mA) was administered through the water. At the end of the fortieth second both light and electrical shock in the lighted end of the apparatus were terminated whether the fish had avoided or escaped the shock. Since the shock was transmitted across the barrier to the opposite side of the shuttle box only to a negligible extent, the response of crossing the barrier provided escape from or avoidance of the shock, depending upon the response latency. After 20 additional seconds the next trial was initiated with onset of light on the side of the shuttle box to which the fish had swum during the first 20 seconds, avoiding the shock, or into which it had escaped.

Fish averaged about 2 avoidances in the first 10 trials, about 4 in the next 10, and 6 on retraining on Day 4. Groups of fish retrained at later times, as much as several weeks after the initial 20 trials, showed the higher avoidance scores characteristic of Trials 21–30 seen on Day 4. Other groups, given Trials 21–30 on Day 1, immediately following Trial 20, performed similarly. The interval between Trial 20 and 21 thus had little or no effect on the total avoidance score for the subsequent block of 10 trials. There was considerable variability in the number of avoidances from fish to fish as well as in the means from group to group. We used a regression equation to predict Day 4 scores for each fish on the basis of performance in the first 20 trials (Davis, Bright, & Agranoff, 1965). The regression was empirically derived from Day 1 and 4 scores of 189 control fish. Predicted avoidance scores for Day 4 for each fish were subtracted from the achieved score as a measure of memory loss. This method of computation aided

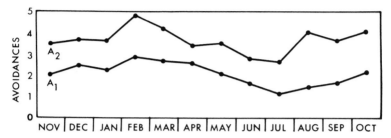

Fig. 1. Seasonal variation in avoidance responding for groups of fish trained in Task 1. A_1 = avoidances on Trials 1–10. A_2 = avoidances on Trials 11–20. Day 4 scores (not shown) were predicted by a regression analysis. (From Agranoff & Davis, 1968.) [Copyright (1968) by the University of Chicago Press.]

in comparing effects on fish differing in absolute levels of avoidance responding within a group as well as in comparing groups of fish that differed in the average number of avoidances during training. Undoubtedly, many factors contribute to this variability. We have observed a distinct seasonal variation in performance. This is reflected in Fig. 1 (Agranoff & Davis, 1968). Even with the regression, avoidance scores were sufficiently low in the summer months to discourage us from running behavioral experiments.

B. EFFECTS OF PUROMYCIN INJECTIONS

1. Task 1

The effect of injection of 170 μg of puromycin dihydrochloride at various times following training of Task 1 are summarized in Table I. This substance was injected intracranially in 10 μl of solution. The fish injected after 1 hour had been returned to home storage tanks following training. Our major findings (Agranoff, Davis, & Brink, 1966) using this task were: (1) Puromycin or AXM when injected immediately following training on Day 1 produced a marked decrease in responding on Day 4. The same amount of antibiotic given 1–2 hours following Trial 20 had no effect. Thus the effects of the injections were temporally related to training, consistent with consolidation hypotheses of memory formation. The injection of saline immediately following training had no effect on memory, while graduated lower dosages of puromycin showed lesser effect. (2) When the agent was injected before training, there were no observable neurological effects as alluded to above, i.e., the swimming behavior of the fish gave no evidence of neuromuscular impairment. Furthermore, there was relatively little effect on Day 1 scores, and yet we found marked memory loss on Day 4. (3) Animals allowed to remain in the training apparatus instead of being returned to home tanks following training remained susceptible to puromycin long after they ordinarily would have "fixed" memory (Davis & Agranoff,

Table I

Task 1: Consolidation

Treatment Day 1	Retention Day 4	N
None	0	72
Puromycin, 0 delay	-2.7^a	36
Puromycin, 1-hour delay	$-.1$	35

[a]Significant at $p < .01$.

1966). Animals remaining in the shuttle box for periods up to 3 hours and not injected showed no memory loss. This striking result suggested that the return to the home tank triggered the onset of the fixation process. (4) Actinomycin D blocked RNA synthesis selectively during the first 2 hours following its intracranial injection. At later times protein synthesis was also blocked. This drug also produced a block in memory (Agranoff, Davis, Casola, & Lim, 1967). These results (Agranoff et al., 1965; Agranoff & Klinger, 1964; Brink et al., 1966; Lim et al., 1970) were interpreted to mean that the formation of long-term memory required ongoing RNA and protein synthesis but that initial acquisition did not. Further discussion of these results and their interpretations are presented below. (5) Electroconvulsive shock was effective over a similar temporal gradient. The period of susceptibility to ECS could be extended by cooling animals from 19 to 9° in the posttrial period (Davis et al., 1965). Cooling alone did not produce amnesia. (6) When puromycin or AXM was given immediately following training, there was no immediate loss of memory. The amnesia developed over a period of 2–3 days.

2. Task 2

Some experiments were performed using shuttle boxes in which fish were trained to swim into the lighted rather than into the darkened side of the shuttle box. This is a more difficult task as judged by avoidance scores (Agranoff & Davis, 1968) but animals could be trained, and memory formation was disrupted with puromycin as in Task 1. Task 3, described below, is also more difficult than Task 1 for the goldfish and has been in use in our laboratory for the past 2 years.

3. Task 3

The shuttle boxes are identical to the ones used in Tasks 1 and 2 except that a clear plastic gate is present over the central barrier and must be deflected by the fish upon crossing. The gate was initially inserted to make the task more difficult in the hope that there would be a greater difference in response scores between Day 1 and Day 4 and also to discourage intertrial crossing. In this task, following 5 minutes of acclimatization in the shuttle box, 20 one-minute trials are given in a single block. A 15-second avoidance period is initiated by a light signal on the side of the shuttle box in which the fish is located. Whether the fish avoids during the first 15 seconds or escapes in the subsequent 20-second shocking period, photodetectors on either side of the barrier register the response and terminate the trial, leaving the fish in darkness until the beginning of the next minute. Twenty such shuttle boxes are interfaced at the present time to a PDP-8 computer which programs the trials and records the responses of each fish on punched tape. Latencies are recorded in tenths of a second. Additional programs facilitate storage of data and calculation of regression equations from groups of control fish run every week. Several thousand fish were studied with the automated apparatus during a period of 1½ years. When control fish perform significantly differently from predicted Day 4 scores on the basis of an annual regression equation,[3] experimental groups for that week are not used. Examination of control groups run in Task 3 (Fig. 2) reveals examples of deviations from the expected response scores of control groups, e.g., the last week in February.

Results of memory studies are in general agreement with our earlier findings with Tasks 1 and 2, but with some interesting differences as follows: (1) In contrast to our results in Task 1, we have found that groups given 170 μg of puromycin 24 hours after training show some apparent sickness in Task 3 as judged by performance on Day 4. Using 130 μg of puromycin in Task 3 injected intracranially in 10 μl (see Table II and Fig. 3), there is no evidence for sickness, although this dose does not produce a complete memory loss. That is, Day 4 scores are not so low as scores for Trials 1–10 on Day 1, but highly significant deficits are seen. (2) Tables I and II show that the "consolidation time" with puromycin in Task 3 is considerably longer than that seen for Task 1, since some disruption of Task 3 is evident if puromycin is injected after 4 hours. No such

---

[3] In Task 3, a separate regression is calculated for two groups of fish, those that demonstrate at least one avoidance response on Day 1 (a) and those that do not (b). Avoidances (A), failures to escape (F), and number of shocks received (S) during Trials 1–10 or 11–20 on Day 1 are designated by the subscripts 1 and 2, respectively. The predicted score for group (b) on retraining, $PA_3 = 5.628 + .129(A_1) + .332(A_2) - .247(F_1 + F_2) - .016(S_1) + .030(S_2)$. For group (a), $PA_3 = 5.082 + .098(F_1) - .339(F_2) - .042(S_1) + .026(S_2)$. The calculations are based on a multiple-regression equation with multiple predictors (McNemar, 1969, p. 197).

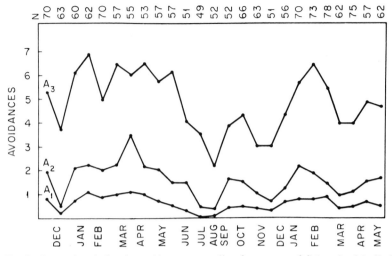

Fig. 2. Seasonal variation in avoidance responding for groups of fish trained in Task 3. A_1 and A_2 are as in Fig. 1. A_3 = avoidances in Trials 21–30 on Day 4.

Table II

Task 3: Consolidation

Treatment Day 1	Retention Day 4	N	Retention Day 7–8	N
Puromycin, 0 delay	−2.96[a]	82	−3.29[a]	36
Puromycin, 4-hour delay	−1.15[a]	80	−1.40[a]	27
Puromycin, 8-hour delay	−.23	53	−.76	24
Puromycin, 24-hour delay[b]			−.04	27

[a] Significant at $p < .01$.

[b] From Davis and Klinger (1969).

deficit is seen for Task 1 after only 1 hour. Animals given puromycin but retrained 7–8 days later show similar deficits as those observed on Day 4.

Although AXM had previously been demonstrated to produce amnesia in Task 1, we used injections of 10 μg of the less potent compound, CXM, in further experiments in Task 3 because of the scarcity of AXM. As can be seen from Fig. 4, this amount of the agent, 50 times the milligram amount of AXM used previously, produced a more intense depression of protein synthesis but of shorter duration (Lim *et al.*, 1970). CXM produced little memory deficit by Day 4 but a degree comparable to that seen with puromycin is obtained upon

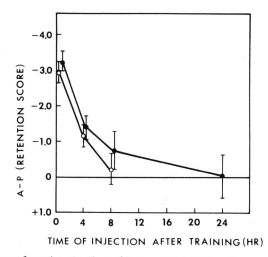

Fig. 3. Effect of varying the time of intracranial injection of puromycin on Day 1 on memory as tested on Day 4 (○) or on Day 8 (●) (some of the fish in the latter group were retrained Day 7). Retention score is obtained as in Fig. 1.

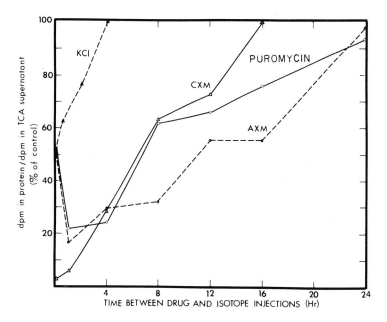

Fig. 4. Effect of intracranial injection of various agents on incorporation of leucine into brain protein (Lim *et al.*, 1970). [Copyright (1970) by Pergamon Press.]

retraining on Day 8 (see Fig. 5). The "short-term memory decay" appears to occur some time after Day 6. This delayed action of antibiotics has been previously observed with AXM (Agranoff, 1970). There may be some evidence of this delayed onset of amnesia in studies in which the various blocking agents were given to fish following 20 minutes in the shuttle boxes (but without training trials) on Day 2, with retesting on Day 8 rather than Day 4 (Davis & Klinger, 1969).

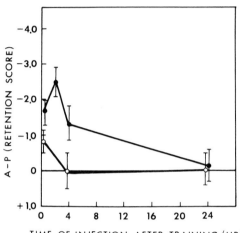

Fig. 5. Effect of varying the time of intracranial injection of 10 μg of cycloheximide on Day 1 on memory as tested on Day 4 (○) or on Day 8 (●). Retention score is obtained as in Fig. 1.

C. The Puromycin Effect and State-Dependent Learning

Experiments were performed to see whether some of the effects of puromycin could be accounted for by "state-dependent" learning (Overton, 1968). For these experiments, the antibiotic was given just before a training session. It can be seen from Table III that puromycin (130 μg in 10 μl) given intracranially both before training and before retraining had no significant effect on the amnesia and that puromycin itself given only before retraining (Day 4) had no significant amnesic effect. The latter experiment suggests that inhibition of protein synthesis by 80% does not affect the retrieval process.

D. Effect of Anesthesia

As discussed below, the effects of blocking agents might be attributed to their nonspecific noxious effects by an interference mechanism. To test this possibility, fish were placed in tanks with 50 mg of Finquel (MS-222) in 1 liter

Table III

Task 3: Dual Injection

Treatment		Retention	N
Day 1	Day 4	Day 4	
Puromycin	—	-2.96^a	82
—	Puromycin	$-.81^b$	28
Puromycin	Puromycin	-3.02^a	20
Puromycin	Saline	-2.50^a	21

[a] Significant at $p < .01$.
[b] Significant at $p < .05$.

of water for 4 hours following training. The antibiotic (130 µg puromycin) was administered as before. If the antibiotic exerts its effect in a heavily sedated animal, it is not likely that any noxious effects of the antibiotic will act as an unconditioned stimulus. It can be seen from Table IV that the anesthetic itself had no effect on retention or on the puromycin-induced amnesia.

Table IV

Task 3: Effect of Finquel

Treatment Day 1	Retention Day 4	N
Finquel, 0 delay	+.04	54
Finquel + Puromycin, 0 delay	-2.71^a	52
Puromycin, 4-hour delay	-1.43^a	46

[a] Significant at $p < .01$.

IV. Possible Mechanisms

A hypothesis consistent with the present results is that normal ongoing RNA and protein synthesis are required for the consolidation of memory. These questions then arise: (1) What is the evidence for the existence of the consolidation process itself? (2) Are the agents reported on here exerting their amnesic effects via their block of macromolecular synthesis or by some other means?

In regard to the existence of a consolidation phenomenon, it should first be noted that this is still an area of scientific controversy (see Chapter 5 by McGaugh). Agents administered at various times after training and causing a decrement in subsequent performance could be doing so not by preventing the formation of memory, but rather by exerting some sort of interference: Behaviorally, by acting as an US to produce conditioned fear or conditioned inhibition; electrically, by generating neuronal noise; or by some other means. Consolidation and interference mechanisms may give very different interpretations to a given experiment. Any evidence of recovery of memory tends in general to argue against the consolidation concept. For example, it has been postulated that electroconvulsive shock does not interfere with memory at all in some circumstances but simply blocks temporarily the performance of a newly learned task. In other instances, it would seem very difficult to distinguish experimentally between interference and consolidation explanations.

How an inhibitor of macromolecular synthesis affects memory is in any event unknown. An agent that inhibits protein synthesis may block memory by preventing the formation of a substance needed for the ultimate expression of memory. It might alternatively cause the formation of faulty proteins or reduce the formation of a particular protein to the extent that other proteins produced before or after the period of inhibition interfere with its expression. The latter is a "chemical interference" of memory.

The present experiments on the effect of antibiotics on the fish behavior tend to support in general the concept of consolidation. Unlike ECS and the convulsant drugs, antibiotics can exert their chemical effects without gross behavioral effects. We can therefore administer the agents before the training session. Such experiments, not previously possible, show that amnesia can be produced as a result of presession treatment. In order to explain such an effect by some mechanism other than by a blocking of consolidation, it becomes necessary to propose backward conditioning in which the agent acts as an US or to suggest proactive effects of the agent on learning. Experiments in which puromycin was used together with Finquel appear to support the idea that the antibiotic is not exerting its effect by acting as a stimulant and also, as discussed below, that it is not acting as a convulsive agent.

Whether or not the antibiotic agents exert their effect on memory by blocking macromolecular synthesis or by some other mechanism is a question which must be answered in order to interpret the biological significance of behavioral experiments in which they have been used. While it is possible that macromolecular synthesis somehow mediates a part of the consolidation phenomenon, it might be via some global aspect of brain metabolism such as respiration (Jones & Banks, 1969). It is also possible that consolidation requires one of the known neurotransmitters, whose concentration is regulated via protein synthesis (Weiner & Rabadjija, 1968). Puromycin has been reported to potentiate the convulsant action of pentylenetetrazol in mice, and we have found this to be true in fish as well (Agranoff, 1970). The possibility that the convulsant property of puromycin contributes to the amnesic effect is diminished by our finding that puromycin aminonucleoside has a similar potentiating effect on convulsions, but not on memory or on protein synthesis.

Since CXM, AXM, and puromycin have all been demonstrated to produce amnesia in the goldfish, and since each of these agents also blocks protein synthesis in the goldfish brain, parsimony at present suggests that the block in protein synthesis is causally related to the memory loss. In comparing the amnesic effects of puromycin and CXM in the present studies, we see that the agents' effects on protein synthesis are over by 24 and 16 hours, respectively. Nevertheless, amnesia may not develop for some days thereafter. In fact, amnesia due to CXM, a shorter acting agent, takes longer to appear than does that produced by puromycin. Verification of a correlation of the inhibition of protein synthesis with the susceptibility to amnesia could be established by injection at various times presession. For example, we expect CXM given 16 hours presession to have no effect on memory. Such studies are presently under way with the use of Task 3.

The interrelationships of the time of injection of an antibiotic, its known duration of chemical effect and the time of retesting the experimental animal have been examined in the hope of giving some insight into the mechanism of the memory-blocking process. It is noted that the effects on protein synthesis of CXM and puromycin are over in less than a day, that the susceptible period is over within a few hours postsession, and that the amnesia may take several days to develop. When CXM is given 4 hours posttrial, little amnesia is found on Day 4 while a significant deficit is detected on Day 8. The diagrammatic representation seen in Fig. 6 is consistent with our findings. We suggest that there is a time (the consolidation period) following training in which there is within the nervous system a strengthening of newly acquired information which proceeds exponentially. The onset of this period for avoidance learning may be triggered by a lowering of arousal accompanying termination of a session and removal to the home tank. On termination of the consolidation period, the strengthening process proceeds linearly until it reaches some final asymptotic level. Examples of how various agents might exhibit their effects are also shown in Fig. 6. While Fig. 6 suggests a theoretical framework that is consistent with our experimental results, it is perhaps also useful to review what it does not tell us. It does not answer the question whether memory requires formation of a specific protein or simply the participation of protein synthesis in some aspect of the memory process. The fact that the agents block memory at all does support a role for a growth process and by implication for a genetic basis for the "engram," the organic basis of memory.

DNA contains the information for the generation of an entire individual including the specification of all the connections in the brain. Within these specifications must also be the capability for changes evoked in the learning experience. It is proposed that the blocking agents prevent this evocation as they

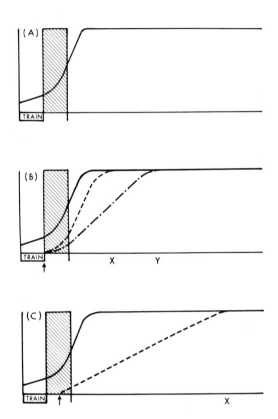

Fig. 6. Diagrammatic representation of effects of disruptive agents on associative strength (ordinate) of recently acquired behavior. During the postsession period (shaded area), an amplification process, presumably synaptic, exponentially increases physiological expression (performance). This process establishes the rate at which strengthening continues until an asymptotic level is reached. (A) Unblocked learning and consolidation. (B) When disruption follows training, a weakening or interfering subtractive process relevant to the training is generated and also increases until it too reaches its asymptotic level. At this time (X) the learned behavior can no longer be elicited. A weaker disruptive agent completes its effect on memory at a later time (Y). (C) When disruption begins after partial fixation, the interference process is subjected to amplification for a shorter period of time. It is therefore expected that an agent, administered some time after the onset of fixation, will ultimately produce a maximal effect on memory but will exhibit a prolonged "short-term memory decay." Long-lasting partial deficits in performance could be produced by interaction of several such processes.

block enzyme induction, cellular differentiation, and other processes under nuclear control.

Ultimately, we would like to understand behavioral changes in the nervous system on a molecular basis. Since the blocking agents appear to act reversibly, a valuable first step might seem to be localization of effects in the fish brain. While such experiments have not yet been performed, it should be remembered that after half a century of searching, electrical and anatomical correlates of learning are still highly speculative. It is too early to say whether the blocking agents will be a more useful tool for localization. It is already apparent, however, that they have given us a valuable tool for studying the memory process—one that separates two distinct phases of memory formation.

REFERENCES

Agranoff, B. W. Recent studies in the stages of memory formation in the goldfish. In W. Byrne (Ed.), *Molecular approaches to memory and learning.* New York: Academic Press, 1969. Pp. 35–39.

Agranoff, B. W. The role of protein synthesis in memory formation in the goldfish. In A. Lajtha (Ed.), *Protein metabolism of the nervous system.* New York: Plenum Press, 1970. Pp. 533–543.

Agranoff, B. W., & Davis, R. E. The use of fishes in studies on memory formation. In D. Ingle (Ed.), *The central nervous system and fish behavior.* Chicago, Ill.: Univ. of Chicago Press, 1968. Pp. 193–201.

Agranoff, B. W., Davis, R. E., & Brink, J. J. Memory fixation in the goldfish. *Proceedings of the National Academy of Sciences of the United States,* 1965, **54**, 788–793.

Agranoff, B. W. Davis, R. E., & Brink, J. J. Chemical studies on memory fixation in goldfish. *Brain Research,* 1966, **1**, 303–309.

Agranoff, B. W., Davis, R. E., Casola, L., & Lim, R. Actinomycin D blocks memory formation of a shock-avoidance in the goldfish. *Science,* 1967, **158**, 1600–1601.

Agranoff, B. W., & Klinger, P. D. Puromycin effect on memory fixation in the goldfish. *Science,* 1964, **146**, 952–953.

Behrend, E. R., & Bitterman, M. E. Avoidance-conditioning in the fish: Further studies of the CS-US interval. *American Journal of Psychology,* 1964, 77, 15–28.

Brink, J. J., Davis, R. E., & Agranoff, B. W. Effects of puromycin, acetoxycycloheximide and actinomycin D on protein synthesis in goldfish brain. *Journal of Neurochemistry,* 1966, **13**, 889–896.

Davis, R. E., & Agranoff, B. W. Stages of memory formation in goldfish: Evidence for an environmental trigger. *Proceedings of the National Academy of Sciences of the United States,* 1966, **55**, 555–559.

Davis, R. E., Bright, P. J., & Agranoff, B. W. Effect of ECS and puromycin on memory in fish. *Journal of Comparative and Physiological Psychology,* 1965, **60**, 162–166.

Davis, R. E., & Klinger, P. D. Environmental control of amnestic effects of various agents in goldfish. *Physiology & Behavior,* 1969, **4**, 269–271.

Jones, C. T., & Banks, P. Inhibition of respiration by puromycin in slices of cerebral cortex. *Journal of Neurochemistry,* 1969, **16**, 825–828.

Lim, R., Brink, J. J.,, & Agranoff, B. W. Further studies on the effects of blocking agents on protein synthesis in goldfish brain. *Journal of Neurochemistry,* 1970, **17,** 1637–1647.

McNemar, Q. *Psychological statistics.* (4th ed.). New York: Wiley, 1969.

Overton, D. Dissociated learning in drug states (State dependent learning). In D. H. Efron (Ed.), *Psychopharmacology: A review of progress.* Washington, D.C.: US Govt. Printing Office, 1968. PHS Publ. No. 1836, pp. 918–930.

Weiner, N., & Rabadjija, M. The regulation of norepinephrine synthesis. Effect of puromycin on the accelerated synthesis of norepinephrine associated with nerve stimulation. *Journal of Pharmacology and Experimental Therapeutics,* 1968, **164,** 103–114.

Chapter 7

SOME ISSUES RELATING ANIMAL MEMORY
TO HUMAN MEMORY[1]

Eugene Winograd

> Let us take a typical case of memory in an animal. Captain Shipp gave an elephant
> a sandwich of cayenne pepper. He then waited for six weeks before again visiting
> the animal, and when he went into the stable began to fondle the elephant, as he
> had previously been accustomed to do. Watching his opportunity, the animal
> filled his trunk with water, and drenched the captain from head to foot.
>
> C. Lloyd Morgan (1909, p. 121)

I. Introduction

In this chapter, I have elected to make some general points about the
relationship between research on animal memory and research on human
memory, rather than to attempt a recapitulation of the contributions contained
in this volume. Following a brief history, I will discuss two issues. The first issue
is the relationship between interference concepts developed in human verbal
learning and their application to animal memory. The second topic I will discuss
is how the approach to memory described generally as "information processing"
may be fruitful in dealing with animal memory.

HISTORY

The major lesson to be gained from scanning the history of animal memory is
that it has never been free of considerations of human memory. A dominant
note was sounded by Aristotle: "Many animals have memory and are capable of
instruction, but no other animal except man can recall the past at will." Since

[1] Preparation of this chapter was aided by Grant GB-18703 from the National Science
Foundation. Discussion of many of the issues with which this chapter is concerned with
Keith N. Clayton, Leo Postman, and D. A. Riley at the Institute of Human Learning,
University of California, Berkeley, and with I. Steele Russell were extremely helpful.
Portions of Section II are adapted from a paper read to the Midwestern Psychological
Association, Chicago, 1969, as part of a symposium on Animal Retention, K. N. Clayton,
chairman.

Aristotle, philosophers have frequently distinguished recollection, the capacity for voluntary recall, from what modern researchers usually call retention. It may be noted that the concept of "search," widely used in current information processing models of memory, is closer to recollection than to retention. I will return to this point later.

In an outline history of comparative psychology, Warden (1927) reviewed some ancient contributions to the problem of animal memory. Warden attributed the following view to Seneca (54 BC–39 AD), "Animals are unable to recall their past experience, their memory capacity being limited to recognition. Thus the horse may recognize a road over which it has traveled before but remembers nothing of it when in the stable afterward (Warden, 1927, p. 77)."

This Aristotelian distinction between man and animal is to be found in the twentieth century as well, in the writings of W. Hunter and E. L. Thorndike, among others. Hunter, in particular, sensed in his distinction between what he called "sensory recognition," on the one hand, and memory, on the other, the difference between a behavioristic and a mentalistic psychology. A great deal of the impetus for Hunter's polemics, best seen in his seminal paper on the delayed response (1913), was a reaction against the anecdotal school of comparative psychology.

As all students of the history of psychology are taught, the impact of evolutionary thought led, in the late nineteenth century, to a search for evidence reflecting the continuity of species. Romanes (1884) represents one attempt to establish this phylogenetic continuity for mental processes. He reported countless illustrations from various sources of what he regarded as mental achievement by animals. Loeb and Morgan both, of course, preceded Hunter in reacting against the anthropomorphism of the anecdotalists. In reading the literature of this period, the modern reader may be surprised to find that memory itself is regarded as a mental process, and thus, following Morgan's canon, must be attributed only when there is no alternative.

Delayed Response

Since Gleitman and Spear (Chapters 1 and 2) represent a return to the investigation of animal memory, but in the context of the tradition of learning and conditioning, it may be instructive to point out the extent of the antimnemonic bias in animal work. In 1913 Hunter introduced the delayed response experiment as a technique for analyzing behavior "where the determining stimulus is absent at the moment of response." Hunter initiated a tradition which has been, in the main, adhered to in the enormous delayed response literature, namely, of viewing the technique apart from the study of memory. As evidence for this statement, one need only read Fletcher's scholarly review (1965) of the delayed response literature and note how infrequently the word "memory" appears.

Hunter, of course, was working at a time when a behavioristic psychology was being founded. One finds in Hunter's introduction to the 1913 monograph an interesting quarrel with Morgan, who had reported on what seemed to him a clear case of memory on the part of his pet dogs. It seems that Morgan's two dogs, on an outing on a mountain, came across "a young coney and they both gave chase. Subsequently, they always hurried on to this spot, and though they never saw another coney there, the reiterated disappointment did not efface the memory of that first chase, or so it seemed (Morgan, 1909, p. 118)." Hunter's comment on this report is, "The obvious criticism of Morgan's illustrations is that they may be simply cases of sensory recognition of the commonest kind (1913, p. 10)."

To the modern reader, the question occurs as to what "sensory recognition" means to Hunter. A clue may be found later on when Hunter discusses whether memory is to be inferred from recognition by an animal of a food bottle encountered earlier. Hunter says, in denying the attribution of memory, "Recognition of this type does not necessarily imply memory or the dating of an experience in one's past (1913, p. 16)." My interpretation of Hunter's attitude is that it stems from his reaction to the widespread anthropomorphism of the day and his zealous application of Morgan's canon. But it must not be overlooked that Hunter shares with the targets of his criticism a view of memory as a mental state involving, it would seem, familiarity, localization in time, and active recall. While the modern reader may, in general, applaud Hunter's rigor and psychological goals, it is unfortunate that the study of memory in animals lost priority.

The early separation between animal learning and what became known later as verbal learning is indicated by the fact that Ebbinghaus had published his work on memory in 1885. The main thrust of Ebbinghaus' effort was to demonstrate that memory could be investigated with the methods of natural science. His orderly functions testified to the success of his insight that memory could be inferred from relearning scores in the laboratory. In a retrospective paper on Ebbinghaus' contributions to the study of human learning and memory, Postman (1968) shows how Ebbinghaus managed to escape the philosophical problems which had made the scientific study of mental processes impossible. Relearning scores may be obtained just as readily from animals as humans in any simple learning task, yet Ebbinghaus' example was influential only within the human learning tradition for many years. It remains for a future historian of psychology to determine whether the reluctance to accept the word "memory" by Hunter and other animal conditioners represents an overly zealous pruning of necessary concepts and the beginnings of a methodological behaviorism which psychology is only now outgrowing.

There is another direction from which the language of the delayed response experiment may be approached. The literature on the delayed response suggests that the theoretical problem has been taken to be "what is learned". Thus, the concern with whether or not representational or ideational factors should be adduced led to emphasis on postural or orienting responses. Experimenters who work on animals have always required an easily defined and measured response. To the student of memory, however, the delayed response setting seems to involve memory for a stimulus. A strong theoretical bias toward response factors, aside from the philosophical status of memory, led to viewing the delayed response setting in nonmnemonic terms. Spear's emphasis on memory as a collection of stored attributes represents a departure from a response oriented theory. In general, one notes in this volume the effects of the behaviorist revolution in the care taken to offer strict operational statements of what is meant by "forgetting" and the accompanying hope of avoiding conceptual entanglement. On the other hand, a willingness to extend the range of cognitive capacities heretofore attributed to the rat may be also seen, mainly in the form of terminology based on information processing. More attention will be given to this trend in a later section.

II. Interference Theory and Animal Memory

The contributions to this volume by Gleitman and Spear represent modern attempts to discuss problems of memory presented by animal data in terms familiar to all students of memory, whether steeped in the Ebbinghausian tradition of verbal learning or in problems of the animal conditioning laboratory. A considerable debt is owed by all of us to Gleitman for his program of research, so fortunately integrated into a single presentation in this volume. His point, made at the outset of his research undertaking, was commendable in its aptness and simplicity. The version of interference theory integrating the vast amounts of data gathered with the verbal learning studies of retroactive inhibition (RI) and proactive inhibition (PI), best represented by Postman's 1961 statement, borrowed to a considerable extent the language of animal conditioning. For instance, unlearning was considered to have "extinction-like" properties, and unlearned associations were supposed to show "spontaneous recovery." Yet, while the flavor of these concepts is that of animal learning and conditioning, they were applied freely to encompass phenomena observed with instructed human subjects who learned, remembered, and seemed to forget verbal items. This development is testimony to the centrality of the language of conditioning theory during the modern era. Although the 1960's may come to be viewed in hindsight as the decade in which the S-R conditioning model was seriously challenged by competing systems or "paradigms" (Kuhn, 1962), the 1940's and 1950's saw the descendants of Ebbinghaus undeniably influenced by the animal learning tradition. While one might argue that this borrowing was merely lip service paid to a reigning point of view, it is undeniable that Irion, in his revision of McGeoch's (1942) standard work on human learning and memory, saw fit to add a chapter on "Conditioned Response Learning" which surveyed the basic phenomena of Pavlovian and instrumental conditioning in animals (McGeoch & Irion, 1952).

At the same time, as Gleitman makes clear, animal learners were simply not interested in forgetting. Historically, it is interesting to note that it was not always so. Some of the earliest American studies on retroactive inhibition were done with rats and little distinction was made for many years between data on retroactive inhibition obtained from human and infrahuman organisms. In 1917, Webb published a monograph on transfer and retroaction in both humans and rats, while Brockbank, in 1919, found in an experiment with rats that interpolation of a rope-ladder problem did not interfere with retention of performance in a circular maze. [I cannot resist including the following methodological point made by Brockbank. He states that during the retention period, his rats were allowed daily exercise in a runway 900 cm long "to avoid the evils that arise from allowing the rats to remain inactive during the retention period (1919, p. 10)."] I think it is safe to say that the relatively few studies on interference phenomena using animals had little effect on the development of interference theory.

Gleitman accomplished at least two things in this context. First, he stimulated others into research on animal memory by pointing out the lack of evidence for the widespread assumption of the durability of learning. Second, in analyzing the constructs of interference theory with the rigor and concreteness required to translate them into actual operations with nonverbal subjects, he forced reexamination of their meaning and utility within the verbal learning domain to which they had been applied. As a consequence, serious questions were raised about the unity of the two domains, and it is to this matter that I wish to address myself.

I will argue here that attempting to apply the concepts and paradigms which have developed in verbal learning to the animal laboratory in order to validate them in their application to verbal learning is unlikely to be successful or even fruitful. While the concepts which are used in verbal learning have their parentage in the animal laboratory, they have altered meanings in the human situations to which they have been applied. This argument is clearly not directed at either Gleitman or Spear, since the directions in which they have moved will be used to support my point.

For the sake of analysis, consider a typical procedure used to investigate RI or PI with rats, the discrimination reversal. Studies representative of work done with this general procedure are those of Crowder (1967), Gleitman and Jung (1963), Koppenaal and Jagoda (1968), and Rickard (1965). In the example used here, the experimental situation is a T-maze with one choice point. The rats in the experimental, or interference, group learn two tasks. Task 1 is learning to go left for a food pellet reward. Upon reaching some performance criterion, say 10 correct left turns in succession, Task 2 is introduced. Task 2 is learning to turn right at the choice point for a food pellet, and training is continued to the same criterion of 10 successes in a row. Note that going left, correct on Task 1, is incorrect on Task 2 and must be extinguished before the rat will learn to go right. The rats in the control group learn only one task, of course.

Imagine that you are the rat in this study learning Task 2, the discrimination reversal. In what ways are your experiences comparable to those of a human subject learning a second list of paired-associates conforming to the basic A–B, A–C paradigm of interference, i.e., same stimuli but different responses across lists? One similarity across situations is that the stimulus conditions remain the same from Task 1 to Task 2. The maze for the rat is unchanged, as are the stimulus terms for the human subject. On the response side, things are not so clear. In the case of the animal, extinction of turning left makes sense, since he will continue to turn left for some time. In the human case, "unlearning" rather than extinction is used to describe what is happening to Task 1 responses during learning of Task 2 since there is an uneasiness about whether extinction operations are applicable. The reason for the uneasiness is simply that human subjects do not make responses appropriate to Task 1 while learning Task 2. In fact, this observation has been the basis for talking about a "selector mechanism" which operates to restrict responding to a given set (Underwood & Schulz, 1960). Perhaps counter-conditioning represents the human case better than extinction.

To one grounded in the tenets of classical interference theory, it is crucial to identify the similarity relationship between tasks. The A–B, A–C relationship noted above has remained the prime interference paradigm. It is not clear to me that the discrimination reversal task I have described for a rat represents an A–B, A–C relationship. Why should it not be taken as representing, instead, an instance of an A–B, A–B_r interlist relationship, where the tasks differ only in the pairings of stimuli and responses, rather than in the replacement of one response by another? Alternatively, I suggest that it can just as reasonably be maintained that the discrimination reversal task is analogous to what is known as "verbal discrimination learning." In verbal discrimination learning, a subject is presented with two words, one of which is arbitrarily designated as correct by the experimenter. The task is to learn which word is correct for each of the many pairs comprising the list. One can readily reverse the contingencies so that the formerly correct choice is now incorrect. Since it is not clear which of the above task analogies is the most compelling one, one does not know which effects to expect since the different procedures do not produce identical patterns of interference (McGovern, 1964). Note, as well, that the human subject would be faced with a list of many discriminations while the rat has difficulty enough in mastering a single choice point. The important role of intratask interference in human learning in itself raises doubts about the possibilities of identical processes being at work.

I would like to turn now to the more significant part of the experiment, the retention test, and raise a different point also noted by Spear. Our animal has first learned to go left at the choice point and then learned to go right. Now, after an interval which has varied in this literature from 1 minute to about 60 days, he is placed back into the maze, runs to the choice point and is faced with a choice: left or right? What should he do? What would any of us do with this information? Which response reflects remembering—which response reflects forgetting?

The point I wish to make with the above example, which represents a great many animal memory studies, is that the rat is uninstructed, at least on the first retention trial. On the other hand, the human subject who has learned two lists *is* instructed as to whether information from List 1, List 2, or both lists is demanded. While to the rat, demand characteristics of the experiment are unspecified, the experimenter has made a decision about scoring. If the study is concerned with PI, the reference response for correct recall is going right, the more recent task. Turning left, to the experimenter, represents forgetting due to interference. By the same token, if the rat happens to be assigned to a retroactive interference condition, Task 1 is the reference task, and turning left represents retention (see Crowder, 1967, for an example of such scoring).

From the point of view of the rat, to talk of "forgetting" from such data violates my sense of parsimony. In line with Spear's data on "lapses," I suggest that data obtained from such studies be regarded as data on changes in response dominance over time. The typical finding, as the contributions of Spear and Gleitman clearly show, is that the dominance of Task 2 declines with time. In an unchanged situation, animals do what they did last when tested immediately; with increasing retention intervals, they tend to perform the response appropriate to Task 1. These are data of basic importance for the student of behavior, since human subjects show much the same pattern (Briggs, 1954).

At least two accounts of this decline in recency over time have been offered. One is simply that the strength of the Task 2 response decays with time until it is equal to the Task 1 response. Gleitman offers a similar argument in his contribution. The reverse side of this point of view is to be found in the doctrine of spontaneous recovery, which states that the change in strength is for the earlier response, Task 1. Task 1 has been weakened during Task 2 learning, but somehow recovers in strength with time. For reasons I have never understood, interference theorists who would not tolerate decay theories of memory had no qualms about positing an equally unexplained increase in response strength over time.

A. Spontaneous Recovery

The spontaneous recovery doctrine has occupied an important role in interference theory, at least until recently, and the overall failure to find empirical support for it (Koppenaal, 1963; Slamecka, 1966) has been a major reason for recent revisions in the theory. It is an illuminating fact that spontaneous recovery has occupied different roles in the animal learning and verbal learning literatures. In animal learning, it has been an empirical phenomenon requiring theoretical explanation; in verbal learning, it has been a hypothetical process invoked to explain the existence of empirical phenomena, such as the growth of PI with time. On the one hand, it has been a reliable phenomenon to be explicated; on the other hand, an unreliable phenomenon, at least until recently (see Postman, Stark, & Fraser, 1968), used to account for something else. As Gleitman states, even if spontaneous recovery were to be found where theory predicts, it would still have to be accounted for.

Spear, quite appropriately, brings Estes' (1955) stimulus fluctuation model into his chapter. This model is indeed an elegant achievement. Part of its elegance derives from Estes' insight that spontaneous recovery cannot be used to explain forgetting, nor can forgetting be used to explain spontaneous recovery. (In fact, the spontaneous recovery paradigm in animal learning is identical to the one used in the foregoing discussion as an example of work on interference. If Task 1 is taken to be conditioning and Task 2 to be extinction, the decline of recency—behavior appropriate to extinction—over time is clearly what the spontaneous recovery literature is about.) Estes attempted to handle all regression phenomena in terms of a single process, namely stimulus fluctuation over time. Gleitman's series of studies in which an attempt is made to come to grips with experimentally controlling stimulus change demonstrates how difficult the problem is. So far, attempts to gain satisfactory experimental control over stimulus change have been unsuccessful. Those who are biologically oriented may take this as supporting the hypothesis that internal change is implicated. The question remains a puzzle at present.

B. Recency and Time

Two different ways of accounting for the observed changes in response dominance over time following mastery of two tasks will be suggested in this section. They are offered chiefly as a demonstration that the possibility exists that not all performance changes over time necessarily reflect mnemonic changes, i.e., a loss in the availability or accessibility of stored information. Let us return to the example of the rat at the choice point of the maze at the time of the retention test (anywhere from a day to 2 months after completion of Task 2 learning). First, I wish to pursue further the decision aspect of his predicament. It should be noted again that the rat is not instructed at this point as a human subject might be ("write down the words from the second list you learned"). May we not assume that, in a changing world, it is adaptive for animals to learn that contingencies change in correlation with the passage of time? As time passes, it becomes less and less likely that things are as they used to be. A creature taking this into account should be less likely to exhibit recency, i.e., do what he did last in that situation, as the "retention interval" increases. In this sense, changes in behavior (or variability, but not in the sense of statistical noise) may be regarded as both adaptive and systematic in a world where change is the rule. Apart from this generalized tendency for "trial and error" behavior to increase over time, the discrimination reversal experiment provides an object lesson in the inconsistency of the environment, since the correct side has already been changed once following a temporal interval.

Another way to approach the observed decline in recency, as suggested by both Gleitman and Spear, is as a failure of discrimination. Let us now assume that rats and pigeons follow the rule: In an unchanged situation, do what you successfully did last. The problem then becomes discriminating what was last from what was first. The longer the interval, the poorer such a discrimination, with failures of temporal discrimination manifesting themselves as performance of the first response, or, to the experimenter, as proactive inhibition. This is a very promising approach, I believe, and has the effect of looking at mnemonic data in psychophysical terms, where the stimulus dimension is time. I suggest that we need more basic parametric data on the effects of such manipulations as varying Task 1 and Task 2 frequency, and particularly, the effects of varying the interval between Task 1 and Task 2. Since the work of Yntema and Trask (1963) on the discrimination of recency with human subjects, a growing body of data on the effects of such manipulations in verbal learning situations has emerged (cf. Hinrichs & Buschke, 1968; Hintzman, 1969; Peterson, 1967). The kind of task described by Capaldi would lend itself to research on recency, since the rat must behave according to what he did last. In this connection, I would be surprised if looking at time relatively, as proportions, did not prove more reasonable than looking at absolute elapsed time. That is, with two tasks there are three time intervals: the time between Task 1 and Task 2; the time between Task 1 and recall; and the time between Task 2 and recall. Stating the first interval relative to the entire interval (Task 1–Task 2/Task 1–retention test) could summarize many temporal manipulations in a single metric of the relative recency of Task 2 and seems to make psychological sense as well. Along these lines, Capaldi's suggestion about the logarithmic nature of the scale for the discrimination of numbers of successive nonrewards is worth further attention.

C. LAPSES AND LOSSES

There are some other points to be made concerning the methodology of animal memory studies as compared to human studies. One observation concerns the distinction made between "lapses" and "losses" by Spear. This is translatable, to a student of human learning, into the measurement of retention by the first retention trial alone, traditionally called the recall trial, or by relearning, the number of trials to regain criterion performance on the reference task. Gleitman and Spear report data from discrimination and avoidance tasks showing losses over time reflected in relearning scores. These persisting effects, particularly when Task 2 is reinforced (PI tests), may pose a problem for simple dominance notions as proposed above. One might assume that the animal, once he has been reinforced during the retention test for the demanded response, has in a sense been instructed. If we take such reinforcement as instructions, then the animal should exhibit criterion behavior thenceforth, given no memory loss. Even this assumption may be anthropomorphic. As argued above, from the point of view of the animal, after a break in the sessions (the interval following Task 1 learning) the rules change. Thus, following the retention interval, the reinforcement rule should have changed again. Why should a single reinforcement disabuse him of this notion?

It is interesting as well to contrast the persistence of a simple choice discrimination over time (where no interpolated reversal learning has occurred) with the evidence for loss when rate measures for "go, no-go behavior" are examined. Granted the difference in sensitivity of measurement for continuous versus dichotomous measures, still the following point comes to mind. I will illustrate it with a hypothetical example. Let us say that I have returned to my boyhood neighborhood after an absence of many years and am assigned the task of retracing my daily walk from home to elementary school. The omnipotent psychologist testing me records my speed of locomotion as well as any errors I commit at the several streetcorner choice points (fortunately, he has such data for the last time I performed this routine years ago). The data would probably show that, while I made no errors at the choice points, I walked significantly slower while exhibiting much orienting behavior. Does this change in behavior represent a loss of stored information? Again, certain changes in behavior correlated with the passage of time away from the situation seem to be reasonable and even adaptive. Perhaps the point is that we need to validate our response measures in the context of a more explicit theory of memory.

D. RECALL, RECOGNITION, OR MOTOR SKILLS

Another point to be raised, which grows in importance with an alternative theory of memory to be discussed later, is whether the kinds of data presented by all the contributors to this volume are analogous to recall or recognition measures obtained with human subjects. I do not propose to resolve this, as I suspect it presents a conceptual impasse. However, given the traditional distinctions made between findings based on recall and recognition data, it is intriguing that the question is so inappropriate with animals. In fact, it may be that a more appropriate literature on human memory than the verbal learning literature, for the student of animal memory, is the investigation of the retention of motor skills. It appears that decay theory has found motor tasks to be a rather fertile ground (see Adams, 1967, Chapter 8, for a review). This observation highlights once more the inadequacy of identifying cognate operations across the animal and human memory literatures. For historical reasons, interference theory developed out of the human verbal learning laboratory rather than from the investigation of motor skills. The separation of these two domains within human memory indicates how wide is the gap between studies of animal and human memory.

It should be clear from the direction this discussion has taken that I regard the demand characteristics of the memory experiment as extremely important. Instructions employed with human subjects are easily taken for granted until we consider probing the memory of a nonverbal organism. One interesting step taken in the direction of removing the ambiguity of animal work in this regard is the addition of contextual cues, as Spear reports in some of his work. Conditional discriminations, or matching-to-sample tasks, have been in the discrimination learning armamentarium since Lashley and may have considerable analytic power when applied to the study of memory. I think that the identity of stimuli in Task 1 and Task 2 so frequently found in animal memory research stems from acceptance of the verbal learning paradigms. At this stage, identity of stimuli across tasks with animal subjects may not be as powerful a procedure as changing them [see recent work by Honig (1969) for an interesting example of task dissimilarity].

E. Recent Changes in Interference Theory

The trend of current research makes it unlikely that a single theory of memory, and certainly not interference theory as it has been known (Keppel, 1968; Postman, 1961), will be found to fit two different data domains. The time seems to be past when a theory of behavior based on learning and conditioning can occupy a central place for students of behavior interested in different problem areas. Within classical interference theory, a new direction is being sought with little reliance on the conceptual terms of animal learning. Spear refers in his contribution to Underwood's (1969) recent emphasis on an attribute model of memory, rather than a model based on associative strength. Postman's recent work (Postman & Stark, 1969; Postman et al., 1968) contains a major reworking of his integration of less than ten years ago. It is of interest in the context of the present discussion that Postman's current theorizing makes spontaneous recovery simply a somewhat trivial outcome of the dissipation of recency, rather than a cause of PI.

Obviously, there will continue to be points of contact between interference theorists, no matter what their experimental organisms are, and theoretical convergence is a goal shared by all. However, the present state of interference theory is not particularly robust, in the sense of general agreement, and caution is indicated in too ready a borrowing of concepts from one domain by those who are working in another.

F. Advantages of Animal Research

The traditional advantages of working with animals still remain, and some of these advantages become particularly clear when memory is involved. For instance, the stress on extraexperimental sources of interference in accounting for forgetting in control groups points to the power of being able to control all facets of an organism's environment. The recent work of Campbell (1967) and Spear (Chapter 2) on forgetting as a function of age is a prime example of this advantage of animal subjects. Finally, I believe that those who study memory in animals are likely to end up raising problems and discovering phenomena which were not encountered in traditional interference research with human subjects and verbal materials. This is likely to occur sooner if, following the example of Gleitman and Spear, animal investigators break out of the biases set when one tries to "test interference theory with animals." Granted that Gleitman's early work started in the direction of testing an interference theory based on phenomena of verbal learning, it is clear that he has ventured in new and exciting directions. For example, his work with goldfish, while exceedingly difficult, is quite a departure from any PI study. Similarly, Spear's work on the Kamin effect in avoidance behavior opens new directions for research. At the same time, his insight into the power of transfer designs for analyzing processes testifies to the advantages of cross-fertilization.

III. Information Processing Models and Animal Memory

Among the contributors to this volume, McGaugh most explicitly uses the language of information processing. His model is clearly a process model, without identification of process with biological structures or functions at this stage, although some possibilities in that direction are contemplated. As such, his model is similar in flavor to recent models put forth to handle human memory, mainly for verbal events, by Atkinson and Shiffrin (1968), Norman (1968), and others (see the recent volume edited by Norman, 1970). These models have in common the postulation of at least two stores, with rules governing the transfer of information from one store to the other. Consolidation theories are obviously intimately concerned with the lability of information during a critical period when the information is not yet in long-term memory.

Since the previous section stressed the borrowing of concepts from animal learning theory by workers in human memory, it is appropriate to mention here that this has been a two-way street. The consolidation hypothesis has its parentage in the work of Müller and Pilzecker (1900) on retroactive inhibition in verbal learning experiments. Furthermore, Tulving (1969) has recently developed a simple procedure with the free recall task for producing retrograde amnesia effects without any physiological intervention as the amnesic event. Essentially, Tulving showed that the insertion of what he terms a "high priority event" into an otherwise homogeneous list selectively lowered recall for the prior item. More specifically, lists of ordinary English words occasionally contained a single famous name (e.g., Shakespeare). At fast input rates, the word prior to the famous name was recalled less often than in control lists not containing the famous name. It will be interesting to observe whether those who do ECS experiments with rats and mice will accept Tulving's procedure as analogous to their own work.

Information processing theories are best distinguished from earlier S–R association theories by their language. Words such as "information," "storage," "retrieval," "buffer," and "search" indicate that the computer has had its effect on students of animate systems. Application of such constructs to the animal conditioning laboratory has already been made by Russell (1966), by Weiskrantz (1966) as described by McGaugh in this volume, and by Estes (1969) and Kamin (1969), among others. When a careful animal conditioner such as Kamin finds it helpful to talk of rats performing a backward scanning process following a surprising unconditioned stimulus, an orienting response by the rest of us is called for. I would like to point out some consequences of using this language rather than the language of learning theory.

One salutary consequence is an emphasis on the continuity of behavior, or information processing, and thereby the avoidance of some problems stemming from a necessarily artificial distinction between learning and memory. While lip service has always been paid to the inextricable relationship between learning and memory, they became isolated experimental literatures in practice. It has been pointed out by many thinkers that, without memory, every interaction with a stimulus would be functionally the first, and hence no learning could be observed. By the same token, it has always been agreed that nothing can be remembered if it is not first learned (encoded or stored). A modern statement of this indisputable point is that of Russell: "It is clear that information storage is an essential prerequisite for a change in behavior through learning. If the salient characteristic of learning is the modification of behavior through experience, then this can be achieved only by means of memory storage. In fact, the phenomena of memory and learning are inseparable: memory without learning is no more feasible than is learning without memory (1966, p. 12)."

To emphasize this continuity of processing theoretically is not to deny that some experimental manipulations are directed toward examination of acquisition processes while other sets of manipulations have as their goal the investigation of the durability or transience of behavior. Yet, at the same time, it may be possible to avoid some problems common to all investigators of animal memory, and represented here in the contributions of Capaldi and Revusky, if learning and memory are not rigidly compartmentalized. For instance, Capaldi finds that a major inadequacy of the learning mechanisms usually postulated in order to handle sequential phenomena is that these mechanisms act too gradually. As an alternative, he offers a conception based on memory, "In contrast to r_g, memories are not learned; that is too conservative a mechanism for altering the animal's internal stimulus environment (p. 119, Chapter 3)." This formulation, I believe, attempts to handle the issue of whether information is acquired gradually or suddenly ("all-or-none") by redefining "learning" as "memory". However, it does not eliminate the problem of describing the laws governing either. Capaldi's chapter demonstrates a further problem arising from the choice of a theoretical language in that it regards memories as internal stimuli.

On the other hand, not only do Capaldi's criticisms of extant learning theories carry considerable force, but he has earned our respect by attempting a major revision of the conceptual focus of a well-established area of investigation. My suggestion here is simply that older concepts may become unduly awkward when confronted with a far-reaching change in outlook. Of course, it is presumptuous to expect any scientist to suddenly change his comfortable old theoretical clothing for the newly fashionable dress, and only reasonable for him to demand evidence that the change is worth the trouble.

Revusky's chapter raises problems stemming from reliance on a conditioning language in a different way. It is clear that he and Garcia have found in the toxicosis problem a difficult puzzle for the student of conditioning. It seems to me that the fact that toxicosis produces aversion to food ingested many hours ago raises a general question of fundamental importance to all students of behavior. After all, temporal succession has been a basic condition of association formation for a very long time. I would like to use a hypothetical example in order to put the phenomenon in more general terms.

Let us imagine that I emerge from my house in the morning and find a flat tire on my car. It occurs to me immediately that at around nine o'clock the previous evening, while driving home, I heard a disturbingly loud noise as I drove over something in the road. Now, 12 hours later, I "associate" the flat tire with the impact. This is not an association in the usual S-R contiguity sense; rather, I have related two events which were separated by a long period of time. I can do this only if I have a record of the earlier event, or memory. But the essential psychological fact here, I wish to argue, is not one of memory. In fact, I have many memories of previous events, and that is the problem. The question to be dealt with is one of trace selection or contact, of how I have related these two particular events. To say, as Revusky does, that two events are related because one has "high associative strength" for the other, or because they are examples of a "relevance rule," may be all that can be said at this stage. However, it is important to recognize that this is not satisfactory. One can see how evolutionary considerations might lead one to expect a channeling of food-related inputs to the same address in storage, thus facilitating retrieval of earlier food events at the time of toxicosis, but food-related events are only one instance of such relationships being established.

The toxicosis experiment may turn out to be most interesting when regarded as raising a fundamental question about the psychology of knowledge. What goes with what, is one way to put it. The traditional answer of temporal contiguity is inadequate and repetition does not seem especially important either, as Revusky emphasizes. Revusky and Capaldi have evidence that the nature of the events themselves determines how organisms code their environment, and it is not unreasonable to suppose that "innate criteria," in Revusky's words, will differ from species to species. The point I wish to make, however, is that when we depart from toxicosis the problem of associating two widely separated events remains. It is an exceedingly difficult but significant problem, dealing as it does with the classification scheme imposed on the environment by organisms. I do not see how a traditional S-R associationism can hope to cope with the problem, while at the same time, I concede that simply talking about "channeling" and "storage address" is not a solution either. It is arguable, though, that the language of information processing may lead to a better formulation of the problem, if only by allowing us to approach it more flexibly and to see it afresh.

A. SEARCH AND JUDGMENT

Earlier, I noted that information processing models are to be found currently in work on both animal and human memory. There is a point, though, where the animal memory theorists reflect the Aristotelian doctrine concerning the difference between memory in animal and man. Two processes widely attributed to human memory are usually called search and judgment. These terms refer to the active retrieval component of directed recall and imply that subjects search through storage, find more candidates than are appropriate, and edit their outputs. An example for the reader is "name the capitals of the fifty states." Animals, to my knowledge, have not yet been accorded the privilege of having an executive editor rummaging around in a mnemonic file cabinet. They are still denied the power of recollection or voluntary recall (although Kamin's backward scan is a foot in the door).

There are two different lessons which might be drawn from this. One lesson is that those who have adopted the language of information processing need to be more parsimonious. In short, the implausibility of a search model for the rat might be taken as a corrective against overly facile attribution of hypothetical processes to all organisms. The opposite lesson is that, since it is hard to reasonably attribute these processes to animals, information processing at some phylogenetic point involves unique processes and we should be prepared to accommodate different principles for animal and human memory. A way out, of course, is to argue that anything psychologically important can be put into operational terms and then applied to any organism. For the present, it is sufficient to observe that students of animal behavior show an increasing willingness to adopt at least parts of information processing models.

It is pertinent here to repeat an interesting observation made recently by Tulving and Madigan (1970) concerning some differences between information processing theories and the older associationist or habit strength models of verbal learning. For the present discussion, the reader should take "students of verbal learning" in the following quotation as "associationists," and "students of memory" as "information processors." Tulving and Madigan say, "If students of verbal learning are preoccupied with time—temporal contiguity between the stimulus and response is regarded as the most important necessary condition for the development of an association—then students of memory are preoccupied with space: information is placed or laid down *in* the memory store or stores, it can be transferred *from* one *to* another, and retrieval is a search *through* the store (1970, p. 440)."

The preoccupation with locations, real or fictional, and transfer among them, makes such models particularly attractive to physiological psychologists, who naturally seek to identify presumed locations with biological ones.

B. COMPUTER SIMULATION

Computer simulation of behavior may be regarded as a special technique for investigating information processing formulations. While consideration of its use goes far beyond the scope of this volume, Feigenbaum (Kimble, 1965, p. 373) has raised an interesting point in connection with the simulation of serial verbal learning with a discrimination net model. In brief, "forgetting" was observed in such a simulation even though no loss of information from the store was written into the program. Feigenbaum's point seems to me worth considering: even with no destruction or erasure of stored information, a retrieval loss due to interference from other stored information may be indistinguishable from a storage loss. While the logical point that proving storage loss is tantamount to proving the null hypothesis has always been conceded (see Weiskrantz, 1966), Feigenbaum's serendipitous finding with simulated learning registers the point with greater force and presents a continuing problem for consolidation theory.

IV. Conclusion

It is clear that the venerable question of whether memories decay remains with us, although the language in which the question is put may have changed to whether a storage or retrieval interpretation of memory loss is preferable. The polarity represented by Spear, who argues for a retrieval theory of loss, and McGaugh, who views the data as compelling a conclusion of a storage loss, is indeed an old one. Yet, it is enlightening to note that McGaugh's preferred model (see his Fig. 12 on p. 235) shows some curves growing with time rather than declining, whereas curves showing loss are amply provided by Spear and Gleitman. Both observations may be taken to confirm McGaugh's point that we study memory by disturbing it; we get at function by creating conditions for dysfunction. Perhaps the most exciting finding for the eventual understanding of the biological aspects of memory is McGaugh's demonstration that retention may be facilitated by an intervention occurring after performance, the inverse of retrograde amnesia. Again, to a decay theorist, such a demonstration is an example of memory dysfunction; the normal decay process is reversed.

Gleitman's paper registers the difficulties faced by all students of memory when theoretical choices must be made. His work on interference processes leads him to reject a version of interference theory based on conflicting associations, but he fears that the seductive allure of decay theory may lead to a blind alley. Still, Gleitman's work with goldfish shows how far he has moved in the direction of decay theory.

There is one obvious point which needs to be made over and over when theories of forgetting are proposed. That point is simply that organisms may be observed both to conserve and to lose information. Consider how much of the data presented in this volume shows strong resistance to forgetting. An assortment of tasks has been shown to be retained over long periods of time. In fact, the crux of both Capaldi's and Revusky's chapters is that the rat can remember a single significant event remarkably well. Obviously, theories of memory have to specify when loss is to be expected and when conservation is to be expected. No current theory seems capable of this, yet evidence for both loss and conservation may be readily obtained. It is to the credit of interference theorists that they have attempted to deal with the problems posed by the conservation of verbal information over time (Postman, 1963; Underwood, 1957).

Gleitman has pointed out that Skinner (1950), in a paper widely taken as a demonstration of complete retention of a discrimination by a pigeon over several years, actually presented data showing severe decrements in rate. To be redundant, the lesson is that there is ample evidence for those who, for theoretical reasons, wish to emphasize retention, while those who are committed to the study of forgetting will usually find evidence of loss. After all, if a subject forgets 25% of the words on a list he learned 2 days ago, at the same time he has recalled 75% of the list. Which finding is taken to require explanation is itself an interesting datum concerning theoretical predispositions.

One further point may be helpful in disposing of some needless debate in this context. Gleitman notes appropriately at the beginning of his chapter that much of the recent growth of research in memory "has concentrated upon the early stages in the life of the memory trace. The biological basis of trace formation, decay and consolidation of short-term memory, disruptions of the consolidation process . . . (p. 1)." In trying to relate the various points of view reflected in this volume, one needs to bear in mind that while McGaugh and Agranoff are largely concerned with the fate of information during a brief period of time after input, the other contributors are concerned with what we may call well-encoded information. This is reflected both in the length of the training procedures employed prior to manipulation of the retention interval and in the time values used in retention testing. Indeed, Gleitman uses a group tested 24 hours following learning as his immediate retention measure, against which all forgetting is measured. That there need be no quarrel about a storage versus a retrieval theory of loss is evidenced by the class of two-stage models of memory referred to earlier which generally attribute storage loss or decay only to the first process.

Once conclusion is clearly shared by all the contributors to this volume: we are a long way from an integrated theory of animal or human memory. However, the reader of this volume must share this observer's optimism at the robust state of experimentation on animal memory, given how recently modern work began in earnest. The wide variety of promising tasks and procedures developed by the contributors, together with the number of provocative findings reported here, may be taken as evidence that the study of animal memory has come of age.

REFERENCES

Adams, J. A. *Human memory*. New York: McGraw-Hill, 1967.

Atkinson, R. C., & Shiffrin, R. M. Human memory: A proposed system and its control processes. In K. W. Spence and J. T. Spence (Eds.), *The psychology of learning and motivation*. Vol. 2. New York: Academic Press, 1968. Pp. 90–197.

Briggs, G. E. Acquisition, extinction, and recovery functions in retroactive inhibition. *Journal of Experimental Psychology*, 1954, **47**, 285–293.

Brockbank, T. W. Redintegration in the albino rat. A study in retention. *Behavior Monographs*, 1919, **4**, No. 18.

Campbell, B. A. Development studies of learning and motivation in infraprimate mammals. In H. W. Stevenson (Ed.), *Early behavior: Comparative and developmental approaches*. New York: Wiley, 1967.

Crowder, R. G. Proactive and retroactive inhibition in the retention of a T-maze habit in rats. *Journal of Experimental Psychology*, 1967, **74**, 167–171.

Ebbinghaus, H. *Memory*. Leipzig: 1885. Republished: Dover, New York, 1964.

Estes, W. K. Statistical theory of spontaneous recovery and regression. *Psychological Review,* 1955, **62**, 145–154.

Estes, W. K. New perspective on some old issues in association theory. In N. J. Mackintosh and W. K. Honig (Eds.), *Fundamental issues in associative learning.* Halifax: Dalhousie University Press, 1969. Pp. 162–189.

Fletcher, H. J. The delayed-response problem. In A. M. Schrier, H. F. Harlow, and F. Stollnitz (Eds.), *Behavior of nonhuman primates.* Vol. 1. New York: Academic Press, 1965. Pp. 129–165.

Gleitman, H., & Jung, L. Retention in rats: The effect of proactive interference. *Science,* 1963, **142**, 1683–1684.

Hinrichs, J. V., & Buschke, H. Judgment of recency under steady-state conditions. *Journal of Experimental Psychology,* 1968, **78**, 574–579.

Hintzman, D. L. Apparent frequency as a function of frequency and the spacing of repetitions. *Journal of Experimental Psychology,* 1969, **80**, 139–145.

Honig, W. K. Attentional factors governing the slope of the generalization gradient. In R. M. Gilbert and N. S. Sutherland (Eds.), *Animal discrimination learning.* New York: Academic Press, 1969. Pp. 35–62.

Hunter, W. S. The delayed reaction in animals and children. *Behavior Monographs,* 1913, **2**, No. 6.

Kamin, L. J. Selective association and conditioning. In N. J. Mackintosh and W. K. Honig (Eds.), *Fundamental issues in associative learning.* Halifax: Dalhousie University Press, 1969. Pp. 42–64.

Keppel, G. Retroactive and proactive inhibiton. In T. R. Dixon and D. L. Horton (Eds.), *Verbal behavior and general behavior theory.* Englewood Cliffs, N.J.: Prentice-Hall, 1968. Pp. 172–213.

Kimble, D. P. (Ed.) *Learning, remembering, and forgetting.* Vol. I. *The anatomy of memory.* Palo Alto, Calif.: Science and Behavior, 1965.

Koppenaal, R. J. Time changes in the strengths of A-B, A-C lists: Spontaneous recovery? *Journal of Verbal Learning and Verbal Behavior,* 1963, **2**, 310–319.

Koppenaal, R. J., & Jagoda, E. Proactive inhibition of a maze position habit. *Journal of Experimental Psychology,* 1968, **76**, 664–668.

Kuhn, T. S. *The structure of scientific revolutions.* Chicago, Ill.: Univ. of Chicago Press, 1962.

McGeoch, J. A. *The psychology of human learning.* New York: Longmans, Green, 1942.

McGeoch, J. A., & Irion, A. L. *The psychology of human learning.* New York: McKay, 1952.

McGovern, J. B. Extinction of associations in four transfer paradigms. *Psychological Monographs,* 1964, 78 (Whole No. 593).

Morgan, C. L. *An introduction to comparative psychology.* New York: Scribner's, 1909.

Müller, G. E., & Pilzecker, A. Experimentelle Beiträge zur Lehre vom Gedächtnis. *Zeitschrift fur Psychologie,* 1900, **1**, 1–300.

Norman, D. A. Toward a theory of memory and attention. *Psychological Review,* 1968, **75**, 522–536.

Norman, D. A. (Ed.) *Models of human memory.* New York: Academic Press, 1970.

Peterson, L. R. Search and judgment in memory. In B. Kleinmuntz (Ed.), *Concepts and the structure of memory.* New York: Wiley, 1967. Pp. 153–180.

Postman, L. The present status of interference theory. In C. N. Cofer (Ed.), *Verbal learning and verbal behavior.* New York: McGraw-Hill, 1961. Pp. 152–179.

Postman, L. Does interference theory predict too much forgetting? *Journal of Verbal Learning and Verbal Behavior,* 1963, **2**, 40–48.

Postman, L. Hermann Ebbinghaus. *American Psychologist,* 1968, **23,** 149–157.

Postman, L., & Stark, K. Role of response availability in transfer and interference. *Journal of Experimental Psychology,* 1969, **79,** 168–177.

Postman, L., Stark, K., & Fraser, J. Temporal changes in interference. *Journal of Verbal Learning and Verbal Behavior,* 1968, **7,** 672–694.

Rickard, S. Proactive inhibition involving maze habits. *Psychonomic Science,* 1965, **2,** 401–402.

Romanes, G. J. *Mental evolution in animals.* New York: Appleton, 1884.

Russell, I. S. Animal learning and memory. In D. Richter (Ed.), *Aspects of learning and memory.* New York: Basic Books, 1966. Pp. 121–171.

Skinner, B. F. Are theories of learning necessary? *Psychological Review,* 1950, **57,** 193–216.

Slamecka, N. J. Supplementary report: A search for spontaneous recovery of verbal associations. *Journal of Verbal Learning and Verbal Behavior,* 1966, **5,** 205–207.

Tulving, E. Retrograde amnesia in free recall. *Science,* 1969, **164,** 88–90.

Tulving, E., & Madigan, S. A. Memory and verbal learning. *Annual Review of Psychology,* 1970, **21,** 437–484.

Underwood, B. J. Interference and forgetting. *Psychological Review,* 1957, **64,** 49–60.

Underwood, B. J. Attributes of memory. *Psychological Review,* 1969, **76,** 559–573.

Underwood, B. J., & Schulz, R. W. *Meaningfulness and verbal learning.* Philadelphia, Pa.: Lippincott, 1960.

Warden, C. J. The historical development of comparative psychology. *Psychological Review,* 1927, **34,** 57–85.

Webb, L. W. Transfer of training and retroaction. A comparative study. *Psychological Monographs,* 1917, **24** (Whole No. 104).

Weiskrantz, L. Experimental studies of amnesia. In C. W. M. Whitty and O. L. Zangwill (Eds.), *Amnesia.* London and Washington, D.C.: Butterworth, 1966. Pp. 1–35.

Yntema, D. B., & Trask, F. P. Recall as a search process. *Journal of Verbal Learning and Verbal Behavior,* 1963, **2,** 65–74.

AUTHOR INDEX

Numbers in italics refer to the pages on which the complete references are listed.

SUBJECT INDEX